The Otter Field Survey and Monitoring Handbook

Otter footprints (main illustrations) and their loping tracks (upper right) and walking tracks (lower right), by David Wege (davidwegenature.uk).

The Otter Field Survey and Monitoring Handbook

A PRACTICAL GUIDE TO FIELD AND CAMERA-TRAP SURVEYS

Melanie A. Findlay & Patrick J.C. White

Pelagic Publishing | www.pelagicpublishing.com

First published in 2026 by
Pelagic Publishing
20–22 Wenlock Road
London N1 7GU, UK

www.pelagicpublishing.com

The Otter Field Survey and Monitoring Handbook: A Practical Guide to Field and Camera-Trap Surveys

https://doi.org/10.53061/XUHK7711

A CIP record for this book is available from the British Library

ISBN 978-1-78427-635-5 Hbk
ISBN 978-1-78427-498-6 Pbk
ISBN 978-1-78427-499-3 ePub
ISBN 978-1-78427-500-6 PDF

EU Authorised Representative: Easy Access System Europe – Mustamäe tee 50,
10621 Tallinn, Estonia, gpsr.requests@easproject.com

Cover photographs: Otter in water – Andy Rouse/naturepl.com; River Teviot – Catherine Hodgkinson; footprints – Mel Findlay; otters on a camera-trap – Mel Findlay & Roger Ingledew.

Typeset in Palatino LT Std by S4Carlisle Publishing Services, Chennai, India

5 4 3 2 1

Printed and bound by Short Run Press, Exeter EX2 7LW

MIX
Paper | Supporting responsible forestry
FSC www.fsc.org FSC® C014540

Dedications

Mel Findlay: To Roger, Molly and Ash

Pat White: To Lucy, Hattie and Martha

Contents

Foreword ix
Preface xi
Acknowledgements xiii

1. Introduction **1**

1.1 Background 1
1.2 Geographical scope 1
1.3 Existing survey methods 2
1.4 A note on terminology 2
1.5 Legislation and licensing 4
1.6 Types of survey 4
1.7 Disturbance 6
1.8 Health and safety 7
1.9 Book structure 7

2. Field signs **9**

2.1 Introduction 9
2.2 Description of field signs 9
2.3 Finding field evidence of otter 30
2.4 Covering the ground 31
2.5 How useful are field signs to indicate otter activity in site surveys? 34
2.6 Key points 35

3. Resting places and breeding places **36**

3.1 Introduction 36
3.2 Legislation and official guidance 38
3.3 Expectations of otter field surveys 41
3.4 Identifying resting places during walk-over surveys 44
3.5 Identifying breeding places during walk-over surveys 48
3.6 Field characteristics of resting place structures 49
3.7 Finding potential resting places during fieldwork 61
3.8 Invasive surveys 64
3.9 Decision tree for identifying resting places 67
3.10 Case study of a holt destroyed under licence in 2010 68
3.11 Key points 71

4. How camera-traps work **72**

4.1 Introduction 72
4.2 Passive infrared detection 72
4.3 Dimensions of camera-trap fields of view and detection zones 73
4.4 Camera-trap illumination 74
4.5 A story to explain camera-trap detection 75
4.6 Detection probability in otters 76
4.7 Key points 81

5. Deployment of camera-traps for otter surveys **82**

5.1 Introduction 82
5.2 General considerations 82
5.3 How should I deploy my camera-traps at potential holts? 89
5.4 How should I deploy my camera-traps at potential couches? 94
5.5 What is the best pattern and timing to camera-trap a resting place? 95
5.6 Other applications of camera-traps for otters 98
5.7 What to do prior to set-up and during set-up, maintenance visits and take-down 101
5.8 Problematic sites 104
5.9 Dealing with faults during camera-trap surveys 107
5.10 Key points 108

6. Interpreting camera-trap data from potential resting places and other sites **110**

6.1 Introduction 110
6.2 Survey effort 111
6.3 Preparing the footage 112
6.4 Layer 1: activity patterns 113
6.5 Layer 2: otter information 115
6.6 Layer 3: behaviours 124
6.7 Types of activity at resting places, characteristics of natal and maternal resting places 125
6.8 Other scenarios when using camera-traps for otters 128
6.9 Quality control 130
6.10 Reporting, summarising and interpreting your data 134
6.11 Key points 137

Glossary 138
References 141
Index 144

Foreword

I first encountered otters over 50 years ago when studying mink in Devon. I was collecting mink droppings, known as *scats*, to find out what the animals had been eating and someone pointed out that the droppings that were a bit bigger and didn't smell unpleasant were actually otter droppings, known as *spraints*. Not long after that I found an otter spraint outside a tunnel under the roots of a beech tree and concluded that the tunnel probably led to an otter's den, known as a *holt*.

Since then, I have spent a great deal of time looking for signs of otters on waterways around the country. Several national surveys of otter distribution have been carried out in the UK over nearly 50 years by searching for signs of otters. This systematic approach has provided evidence of a recovery in the otter population from being close to extinction in England in the late 1970s to being widespread throughout the UK today.

Sign surveys are good for answering the question 'Are otters here?' but ecological consultants need to answer different sorts of questions. The fact that otter resting and breeding places are protected by law means that key questions include 'Is this an otter resting place?' and 'Do otters breed here?'

This book provides a sound basis for answering these questions. Mel and Pat have carried out long-term, systematic studies of otter activity using camera-traps at actual as well as potential resting and breeding places. This has provided a substantial body of evidence, which they use as a basis for guidance on what information you need to collect before you write a report stating that 'this tunnel, under the roots of a waterside tree, is a resting place' or 'there is no evidence that this area is a breeding place'. I now conclude that the tunnel under the beech tree I found in 1972 might, or might not, have been a resting place!

They also address the issue of confusing terminology. In the first paragraph of this foreword, I used 'scats' and 'spraints' as synonyms for droppings, though for different animals, and 'holt' for an otter's den. The fact that otters live in holts, badgers in setts, and foxes in earths is quite well known – though you do see the occasional Badger's Holt as a house name. Less well known outside the world of enthusiasts is that otters also have *couches*, *hovers* and *haul-outs*, their footprints are known as *seal* and their tails as *rudders*. These terms, together with spraint and holt, were originally used by otter hunters but some have crept into the vocabulary of people who study these animals and carry out surveys. Unfortunately, they have not always been used consistently. Is a hover a resting place or not? Mel and Pat have used the evidence they have collected to address this problem and clearly set out how the terms should be used and, in some cases, when they should not be used.

This book will be a standard reference for ecologists and ecological consultants who need to understand the otter's use of the landscape and find and identify resting and breeding places. It clarifies what signs to look for and how to recognise and interpret them. It also provides a comprehensive introduction to the systematic use of camera-traps to record otter behaviour, how to analyse it and how to interpret it. This not only enables us to give more consistent, evidence-based advice to protect otters, but will also help to improve our understanding of otter ecology and behaviour.

For all these reasons it is a very welcome addition to the literature on otters, and anyone who uses camera-traps for the systematic recording and study of animals will find it a mine of useful information and advice.

Paul Chanin
Independent Mammal Ecologist (retired)

Preface

Mel Findlay

As a relatively recent graduate, way back in the day, I applied for an 'Otters and Rivers' project officer post. I knew nothing about otters and so prepared for the interview by reading Paul Chanin's *The Natural History of Otters* (Chanin 1985). In the interview I must have sounded like I knew a little about otters, but I then admitted that I had never smelled spraint although I had read 'the book' cover to cover. Maybe my honesty (or audacity?) won my interviewers over. I landed the job that nurtured my early fascination with otters.

A move to Scotland to become a survey team supervisor with the Scottish Wildlife Trust enabled me to indulge this interest because otters were recolonising the rivers in Edinburgh. At this time, it was all about spraint, because camera-traps were not available. My career evolved as a freelance surveyor and then setting up a consultancy, with otters remaining a passion in work and in my spare time. Camera-traps eventually became available and I eagerly engaged with all the possibilities they offered. At first, every mouse zooming across the screen and every munching rabbit was a cause of utter delight, and I still get that buzz of excitement from some of the bycatch, such as a posse of female goosanders sitting on the bank and chuntering to each other, or a water rail casually walking past.

The real turning point came when I put my neck on the line with a client and postulated that a structure on a development site was a natal holt and that it required camera-trapping. With hindsight, I think I had more jam than Mr Kipling: the first batch of grainy footage recorded daily movement of a female otter in and out of the structure, and eventually two tiny cubs emerged. What was more, an adult male also rested in the structure at regular intervals while it was a natal holt, and this made me realise that there was a whole lot of interesting discoveries to be made. Roger, my partner, and I diligently continued at the site for five years, logging all of the activity and learning so much about otters and camera-trapping. I realised we had gathered important data but didn't know 'how to get it out there', until I met Pat. Straight away he saw the value of the data and potential for more research. That meeting led to years of more research, which we have published in several journal papers, and now to this book.

Pat White

I was primarily a bird ecologist but in 2014, when I'd not long started out in academia, I was keen to expand my horizons and, somewhat on a whim, attended a fantastic Chartered Institute of Ecology and Environmental Management (CIEEM) training course on otter ecology and surveying run by Mel. Looking back, I think I'd assumed that for such a charismatic, relatively large-bodied mammal we already had all the answers about ecology and survey. However, during discussions with Mel it was clear that there were lots of unanswered questions about otter behaviour and thus how to accurately survey them and identify and study their resting and breeding places.

This was also a time when camera-traps – as we now know them – were in their infancy but were starting to reveal some interesting, previously unseen behaviours and patterns of use at otter holts. Mel was an early adopter of this technology. At the same time as her seeing their value in otter surveying she explained that many questions remained about how to get the most out of this camera-trapping approach, such as placement, timing and interpretation. We decided to pursue this as an avenue of research in about 2015, and this continues to today. We were adamant that we wanted to conduct applied research that would have direct benefit to otters. Our goal was to provide a firmer evidence base for ecologists studying otters, so that there could be more confidence in how we identified and interpreted resting and breeding places, ultimately to inform their protection. As such, as it started to produce useful results we combined this research with direct training for consultants, civil servants and conservation professionals, in articles aimed at ecologists via our professional body CIEEM and, ultimately, this handbook.

Although as UK-based ecologists there is some focus within this book on UK legislation and policy, we hope that many of the field survey and data interpretation approaches will be useful for studying otters across their pan-Eurasian range and that they will help reveal more about the behaviour of this keystone species.

Acknowledgements

Roger Ingledew, thank you for helping to sort and record the tonnage of footage, for your innovation and DIY skills, your clear thinking on technical issues and your support and sense of humour (excluding the dad jokes – for those go directly to dad-joke jail!).

Thank you for the invaluable input to earlier drafts by Paul Chanin, Graham Scholey, Grace Yoxon, Paul Yoxon, Ben Yoxon, Lucy Hughes, Trevor Hughes, Lindsay Mackinlay and Julia Massey. Neil Middleton gave us valuable advice when we were first considering writing a book. Thanks to our publisher Nigel Massen for support, encouragement and the odd deadline extension during the process. Thanks to editors David Hawkins and Cath Harris.

Edinburgh Napier University's School of Applied Science and Centre for Conservation & Restoration Science provided support for the writing of this book. Prof. Rob Briers contributed substantially to some of the underpinning research. Thank you to all the ecologists and other professionals we have worked with while studying otters over the years who have provided valuable input and discussion, and to numerous landowners for access. Derek Gow and The Chestnut Centre allowed access to camera-trap captive Eurasian otters. Thanks to Dr Elizabeth Chadwick of the Cardiff University Otter Project for otter postmortem data.

The Frontispiece illustration and illustration of otter and mink prints were kindly provided by David Wege (davidwegenature.uk) and adapted from his book *Mammal Tracks of Europe*. Images were kindly provided by Steven Parker and Lisa Thomson (Nàdarra Ecology Ltd), Catherine Hodgkinson, Barbara Macfarlane, Julia Massey (M&H Consulting Ltd), Elaine Rainey, and Sara Toule and Anna Redpath (SLR Consulting Ltd).

1. Introduction

1.1 Background

This book aims to provide support for all those undertaking surveys of Eurasian otter *Lutra lutra*. Each chapter builds knowledge to help the reader develop from a novice surveyor who can find and identify spraints, to an accomplished ecologist able to undertake field surveys and interpret them in light of published literature as well as knowing when to follow up field surveys with further survey or monitoring with camera-traps.

The handbook presents the current evidence base and discusses how this can affect surveys, so that the ecologist can take on board any official guidelines and implement them as confidently as possible. For example, most guidelines on site surveys for development require ecologists to locate and identify the resting places of otter, but prior to recent research there was no advice to support this task. Therefore, out of sheer necessity, ecologists ran with 'received wisdom' where there was nothing to support – or refute – the conclusions drawn from fieldwork. This book contributes to filling in missing information, based on the most recent research.

1.2 Geographical scope

This book has been written from a UK perspective with observation of otters in a temperate climate, and it is based on the experience of UK ecologists working on Eurasian otters in the UK. Some of the aspects relate to meeting current legislative requirements that are EU/UK focused. However, much of the content on field signs, behaviour, and adopting survey approaches that build on the best evidence about otter ecology and behaviour have a broader reach. This may raise important considerations for survey in other areas of the Eurasian otter range that are fundamentally different, for example the Mediterranean or the Middle East. Sections on how camera-traps work and how to account for this when deploying them (chapters 4 and 5), while focused on Eurasian otter monitoring, may have broader application beyond Eurasian otters, for other otter species, semi-aquatic mammals or other denning mammals.

1.3 Existing survey methods

The otter survey method was developed in the UK in the 1970s by the then Nature Conservancy Council to facilitate otter monitoring across England, Wales and Scotland (Romanowski and Brzeziñski 1997). It is centred around sampling rivers at bridges, where otter field signs are more likely. Each selected bridge is surveyed for otter field signs, together with 600 m of one of the riverbanks. This method was adopted as a standard approach and is described in 'Monitoring the Otter' by Paul Chanin (Chanin 2003a), which is most often cited as the method for commercial otter surveys. This method is very much aimed at monitoring otters over large spatial scales such as expansive river catchments or entire counties or countries. It was designed so that the favourable conservation status of otter could be assessed within Special Areas of Conservation by collecting field sign and habitat data.

The method requires the catchment to be sampled using survey sites/points at an even density across a given unit area. For instance, in the most recent Scottish survey a sampling density of two sites per 10 km square was used aiming for a minimum of 5 km between survey points (Findlay et al. 2015). Each survey site targets a feature that otters are likely to leave signs at, such as a bridge or a weir. At each survey site, a riverbank or shoreline is walked and field signs of otter and often variables relating to the riparian habitat are recorded. The distance walked in individual monitoring projects varies, with 600 m being the maximum, but some surveys stop when the first field sign of otter is recorded. While this is a valuable reference, and keen surveyors should be familiar with it, it is not a blow-by-blow method of how to undertake a survey of otters, such as at a proposed development site.

It is generally acknowledged that otter surveys for development sites or other projects need to be bespoke and that some aspects can be varied according to the site and the project. It is hoped that the information in this book will precipitate more considered survey approaches, moving away from the practice of a single field visit to record otter field signs, which is often combined with a survey for water vole *Arvicola amphibius*.

1.4 A note on terminology

Understanding terms that describe the type and function of otter resting places is important, and readers should familiarise themselves with the terms and how we use them. Many terms have been employed over the years to describe the place in which an otter rests, such as 'holts', 'couches', 'natal holts', 'breeding holts', 'maternal holts' and 'cub-rearing holts'. In writing this book we fell deep into the terminology rabbit warren, using terms with more than one interpretation and a good number of other muddles that otter terminology led to. To solve this, after much thought, we standardised our terms throughout the book. You may take on board our approach or you may not, but it is important to understand which terms you use, apply them consistently and use a glossary in any reports you produce.

We use 'resting place' as the generic term to describe a place in which an otter has rested. In previous research that we cite in this book, we have used the term

Table 1.1 Terms used to describe the function of an otter resting place, based on Robitaille et al. (2020). A resting place is any place or structure used by one or more otters between moving bouts for resting and/or sleeping (prerequisite).

Term	Definition
Natal	Place or structure in which cubs are born. A female can only use one per breeding cycle and usually occupies this structure for eight–ten weeks.
Maternal	Place or structure used by a reproductive female after the use of a natal site, and where the weaning of the kits/pups generally takes place, estimated to be between five and six months old. A female will use a number of maternal resting places.
Breeding	Generic term for places that encompass natal and/or maternal functions.

'resting site', which is synonymous with resting place, but we adopt the latter here to align with terminology used in legislation. The term 'resting place' does not relate to the type of structure or the reproductive status of the individuals occupying it; it is purely and simply a place where one or more otters have rested and where resting may be by individual otters or by family groups. Types of resting place can be described by their function and also by the type of structure they are in. To describe the function, we borrow from the terminology for the martens *Martes* group (Robitaille et al. 2020) for clarity (Table 1.1).

The otter-specific terms in common parlance that describe the type of structure of the resting place – holt and couch – can be added as further descriptors if appropriate. A holt is an otter's resting place which is enclosed. It could be underground such as in burrow systems in riverbanks or the peatlands of Shetland, in hollow trees or in any other enclosed structure. The term 'couch' is often used to describe resting places that are above ground, such as in dense seasonal vegetation, gorse scrub or on riverbank ledges. It should be noted that the function of a resting place can change over time. For instance, a structure could be used sporadically by a male for a few months, then become a natal holt and subsequently be used as a maternal holt after the cubs have emerged.

The term 'hover' has been adopted more recently by some, but definitions provided by practitioners vary and it is a non-standard term. In times gone by, hunters used this word to describe any place frequented by otter. An additional term that is occasionally used in survey reports is 'haul-out', likely borrowed from sea otter *Enhydra lutris* or seal terminology. Again, accepted definitions of this term in the context of otters can't be found, although it is assumed that these are areas in which an otter can lie by the side of the water while they eat and/or groom. To avoid confusion, the terms 'hover' and 'haul-out site' are not used in this book.

'Temporary resting site' is occasionally used for a purported resting place thought to be used just once for a very brief period. This is neither a helpful term or idea, since such places are usually identified by spraint in association with a level area such as a ledge. Interpreting the duration of rest and/or the likely reuse of a place for resting based on field signs, especially spraint, is not an approach we support. Firstly it is not evidence based (discussed in more detail in Chapter 3), secondly it easily cascades into calling any spraint site in structural habitat a resting place, and finally the nuance of a 'temporary resting site' may lead to confusion in the application of legislation. In this book we present our own data and examine the duration that otters spent at resting places and their frequency of use, and place this information within the ethos of the legislation and guidance.

Where it is not known whether a structure is used for resting, a prefix 'potential' will be used, for example potential resting place, potential holt, potential natal couch.

1.5 Legislation and licensing

The otter was first protected across the UK by inclusion on Schedule 5 of the Wildlife and Countryside Act 1981, which protects individual otters from harm, safeguards their places of rest from damage, and protects otters from disturbance while occupying a place of rest. Although there have been inevitable amendments over the years, the otter is still protected by the Act in England and Wales. In Scotland and Northern Ireland, the otter was removed from Schedule 5 after the adoption of the Conservation (Natural Habitats, &c.) Regulations 1994, commonly referred to as the Habitats Regulations, because the two pieces of legislation are so similar.

Across the EU and the UK, the otter is a European Protected Species by its inclusion in Annex IV of Council Directive 92/43/EEC, commonly referred to as the Habitats Directive. This mandated member states to implement the directive and enshrine certain protective measures for species into domestic law. The UK responded to this by enacting the Habitats Regulations. The otter was included in Schedule 2, which is a shortened list of species listed in Annex IV of the Habitats Directive that occur in the UK. The Habitats Regulations have been amended over the years but still form the 'core' of the otter's strict protection, and follow the ethos of the Habitats Directive. This remains so even after Brexit. Future legislation is likely to change and differ between England, Northern Ireland, Scotland and Wales, and therefore the relevant statutory nature conservation organisations should always be consulted.

When surveying for otters, the key message to bear in mind is that these animals are protected from disturbance when occupying a resting place and that damage to resting places is an offence. Therefore, it is incumbent on the surveyor to understand the type of survey they are undertaking (walk-over, invasive, camera-trap monitoring) and be clear as to whether they require a licence in their country to search for or monitor otter resting places. This varied at the time of writing.

1.6 Types of survey

Focusing on site surveys, for development or to facilitate riparian management of, for example, a nature reserve (and putting presence/absence surveys of large catchments/areas to one side) it is useful to think of a hierarchy of survey methods.

Confirming the presence of otter on site is rarely needed; presence should be assumed across most of the UK where there is suitable habitat. This approach obviates the potential for false negatives (i.e. where there is an absence of field evidence in an area occupied by otters).

Each survey method has different aims and methods and requires additional skills, experience and expertise (Table 1.2).

Other methods are available to directly observe otters at night. Thermal imagers can be used but these need to be mid to high spec to view from a practical distance. Cheaper thermal devices may be good at close distances but are not sensitive enough to find otters at distances much over 15 m. Likewise, infrared night-vision aids are only useful at close quarters, which, for most applications for otter would not be good enough. However, the real issue with using this technology for direct observation is that the conclusions that can be drawn are limited and often do not offer much more than 'presence' of otter, which is better gauged using an indirect technique such as a field sign survey. There is no established approach to follow in terms of survey

Table 1.2 Summary of the types of otter survey and monitoring methods and where to find information in this book

	Survey methods		
Survey aim	**Field sign survey/ walk-over**	**Invasive survey of potential resting places (licence may be required)**	**Camera-trapping potential resting place (licence may be required)**
Confirm presence of otter on site	Chapter 2		
Identify resting places and potential resting places	Chapter 3	Chapter 3	Chapters 5 and 6
Identify whether a structure is a natal resting place			Chapters 5 and 6
Identify whether a structure is used for rearing young cubs, i.e. a maternal resting place			Chapters 5 and 6
Identify whether a breeding female is active on site, with or without cubs			Chapters 5 and 6
Assessing terrestrial routes	Chapter 2		Chapters 5 and 6

effort for aims such as finding resting places or assessing activity or importance of habitat. Finally, the time and cost of one or more surveyors sitting night after night with high-end thermal devices has to be questioned with regard to what can be gained from the data.

The exception to this would be in coastal areas where otters are active day and night, allowing 'regular' equipment such as binoculars and telescopes to be used for direct observation during the daytime. The likelihood of being able to watch otter(s) for long periods of time is far greater because larger expanses of coastline may potentially be viewed by a single surveyor coupled with the home ranges of otter being potentially much smaller. Again, there is no protocol to follow but it would be possible to locate some resting places and observe family groups. This would require a truly bespoke approach, underpinned by a high level of experience and observer skill.

1.7 Disturbance

Surveying otters and monitoring their resting places or natal and maternal sites inevitably brings potential for disturbance, and this must be carefully considered and mitigated when planning and implementing fieldwork. Torches may be used in walk-over surveys to aid the brief inspection of structures to rule out any potential as resting places and this is different from an invasive survey, which has most potential for disturbance. An invasive survey involves a thorough examination of a structure that has promise as a resting place, using a torch, flash photography or extendable camera equipment to look for nests or latrines.

Camera-trapping potential resting places, which requires repeated surveyor visits to a resting place, is an aspect on which we provide detailed guidance in this book. EU guidance (European Commission 2021) on interpretating disturbance in Article 12 states that disturbance is an act that impacts 'the chances of survival, the breeding success or the reproductive ability of a protected species, or that leads to a reduction in the occupied area or to a relocation or displacement of the species'. It should be noted that this does not define disturbance as meaning any observed behavioural response. Of course, we should expect that otters are aware that camera-traps are present, and on camera-trap footage we have observed behaviour such as sniffing the camera-traps, but we have never observed any responses that suggest displacement or impairment to survival or breeding. In fact, at several locations we have observed otters choosing to sleep directly in front of camera-traps, and repeated use of resting or breeding places while camera-traps are in situ. The necessary set-up, maintenance and take-down visits made to camera-trap sites also have potential to cause disturbance, and we provide some advice about minimising time spent on each visit, impact of human presence when there, and the optimal frequency of visits to balance potential for disturbance against risk of data loss.

It is important to ensure you are appropriately licensed during surveys. The need for a licence and the accompanying conditions can vary widely between the four UK issuing bodies, Natural England, Natural Resources Wales, NatureScot and the Northern Ireland Environment Agency (White and Findlay 2024). The licence conditions should enforce sensible working practice, which controls the risk of disturbance. Despite all the licence conditions being followed, it is still possible that an ecologist could disturb an otter but the licence would mean that this would not be unlawful. It is therefore incumbent on the surveyor always to take precautions against any disturbance.

The first consideration when thinking about any type of survey or monitoring is whether it is necessary. Potentially disturbing surveys must provide data where there is a 'need to know' rather than where it would be 'nice to know'. Having established a need to know, a key question must be asked when considering invasive and/or camera-trap surveys – what is the greatest potential risk for otters: (1) the idea that the surveys could disturb otters (something the evidence disputes when strict protocols are followed) or (2) the risk that poor survey protocols (through well-meaning precaution) miss key observations that would identify sites correctly as a resting or breeding place, which ultimately results in a protected place being unintentionally disturbed or destroyed?

1.8 Health and safety

The approach to health and safety varies between companies, organisations and institutions. Often, larger companies have more restrictions on what a surveyor can and can't do, for instance disallowing a surveyor to be within 2 m of running or standing water. Many otter surveys abiding by such conditions verge on pointless. Otter surveys intrinsically pose more safety considerations than other types of survey, nearly always relating to the risk of falling into water and pathogens associated with that environment, for example rat urine and sewage discharges. Many of these can be covered within a risk assessment and by implementing certain measures, such as no lone working, having one person on the bank if one person is on the water's edge or wading in the water, the terrestrial person carrying a throw rope, use of a wading stick, use of gloves to avoid contact with bank and associated pathogens, and use of hand sanitiser and covering cuts as standard. A boat may well be needed too, to access areas of shoreline that can't be reached on foot, such as expanses of dense trees or shrubs. Please exercise appropriate caution if handling/sniffing otter faeces.

Health and safety concerns can be mitigated most of the time. If they cannot, this is a major caveat for any survey. However, not undertaking a good survey due to consultancies avoiding risk rather than controlling risk is likely to lead to problems down the line due to having a poor baseline.

1.9 Book structure

The following chapters follow a natural progression of otter survey and monitoring requirements, from field signs that are useful during walk-over surveys (Chapter 2), the nature of resting and breeding places and how to identify confirmed or potential resting places (Chapter 3), how to deploy camera-traps at potential resting places or other features (Chapter 5), and finally how to best interpret camera-trap footage

(Chapter 6). Chapter 4 takes a slight sidestep to describe how camera-traps function and what drives their detection probability, with particular reference to research on otters, so that some of the more practical field aspects of their deployment described in Chapter 5 are well grounded in evidence.

We have tried to include many images to illustrate and support our information. The hope is that this book can be employed both as a reference prior to and after surveys, or taken into the field as a useful guide.

2. Field signs

2.1 Introduction

This chapter provides descriptions of general field signs of otter and help with how to find them. If in doubt about whether you require a survey license, you should check with your regional licensing body.

Resting places, another important field sign, are considered in Chapter 3.

2.2 Description of field signs

2.2.1 Spraints

Spraints are faecal scent marks left by otters at strategic features. They are usually easily identified and a great start to tracking otters. If we use the analogy of 'sausage, mash or gravy' to describe their form, spraints are generally like mash and/or gravy. They tend to have a black matrix but this can be translucent white or a brownish green (Figure 2.1a). They vary in size from a tiny black tarry smear the size of your little fingernail, to a chunky deposit as large as an index finger. Fish bones and scales often give a characteristic spikey surface texture while frog bones tend to look shorter and stumpy in comparison (Figure 2.1b), but crayfish remains, fur, feathers and occasionally fish roe can give a very different appearance. Coastal spraints can be much chunkier, being bulked up by indigestible crustacean remains such as crab (Figure 2.1c). The best and most reliable characteristic that separates spraints from the droppings of other mammals (and birds) is their sweet, musky odour. This is often likened to the smell of newly cut hay or jasmine tea; either way, it is a completely unexpected scent and is unique to otter. This scent can be distinguished even if there are strong fishy overtones, which are often more pronounced with coastal otter spraints. The muskier or more 'ottery' the smell, the greater the likelihood that the depositing individual was an adult (Kean et al. 2015).

Smell is therefore key to an accurate identification. If the spraint lacks any food remains it can appear as a black tarry smear, in which case, as a surveyor, you need to sniff it in situ. If the surveyor lacks a reasonable sense of smell, a small part of a spraint can be scooped off with a leaf or similar, bagged and taken home. The scent will intensify in the bag over a couple of days, which may help those with a poorer sense of smell to identify it. It is also rumoured that placing spraint on a radiator can be quite effective too, unless you object to your house smelling of otter.

Figure 2.1 Variations in colour and form of otter spraints: (a) dried black spraint on top of boulder with greenish slimy spraint further down; (b) 'mash' form containing fragments of short and chunky frog bones with a black and mucoid matrix;

Figure 2.1 (*cont.*) Variations in colour and form of otter spraints: (c) coastal spraint – large, chunky and brown with crustacean remains; (d) very weathered spraint site with the matrix washed out, leaving a few scales;

Figure 2.1 (*cont.*) Variations in colour and form of otter spraints: (e) a variety of spraints on a single spraint site.

Weathered spraints generally become paler as the black matrix gets washed through, leaving a mass of hard remains; the elongated mass of pale grey fragments sometimes resemble cigar ash. As the spraint gets washed away, a few stubborn bones or fish scales can remain glued to a rock (Figure 2.1d). The appearance of sprains thus varies due to weathering, the prey otters feed on at different times of the year, and the habitat. Even fresh spraints can vary in appearance when all things are relatively equal (Figure 2.1e). The surveyor's nose is therefore the preferred tool for confirming spraint.

Mink scats can closely resemble otter spraints in appearance (Figure 2.2a). The contents are of less importance when separating mink scats from spraints because there is overlap in their diet. The presence of feathers and fur should not preclude identification as spraint, nor should an abundance of fish preclude identification as a mink scat. Their scent is often the only way to differentiate the two: inoffensive otter spraints can be sniffed multiple times, whereas one sniff of a mink scat is more than enough if it is fresh! Bird droppings are also commonly encountered and can confuse the beginner. They can superficially resemble spraints but are rarely pure black, often have opaque white sections (Figure 2.2b), and they smell unpleasant. A summary of the key identifying characteristics of otter spraint, mink scat and bird droppings, all of which might be found in otter habitat and cause confusion, is in Table 2.1.

Figure 2.2 Droppings that look like otter spraints but need closer examination and/or to be checked for the scent of otter spraint: (a) mink scat, which is black like a spraint but tends to have more shape and is either smooth as shown or with a twisted form. However, otter spraints can be very similar if they have been eating birds/mammals and the only reliable distinction is the scent; (b) a black and shiny pheasant dropping. Note the small area of opaque white, which indicates bird dropping. (Mink scat photo: Elaine Rainey)

Table 2.1 Comparison of the main characteristics of otter spraints, mink scats and bird droppings.

Characteristic	Otter	Mink	Bird
Colour of matrix	Most are black, occasionally translucent white/cream jelly, sometimes referred to as anal jelly. Can be a combination. Can be brown or dark greenish, but this is relatively uncommon	Black	Black, brown, white (opaque) or combinations thereof. Pheasant droppings and other birds' droppings can be black and glossy (Figure 2.2b)
Form	Mostly mash or gravy, occasional sausage. Can resemble the form of a mink scat if it comprises feathers or fur	Can have a twisted form but can also be smooth (Figure 2.2a) and occasionally mash	Tube-shaped, can be mash or gravy
Texture of matrix	Mucoid	Paste	Paste or granular; can be slimy
Smell	Sweet-smelling, fresh-cut hay, jasmine tea, musky, fishy. Retains scent for a considerable time as it weathers	Repulsive! 'Scent' is lost quite quickly	Various but generally unpleasant

Spraint-sniffing etiquette

A twig or similar can be swiped through the spraint and sniffed. This avoids any hygiene risks from using hands, avoids any deposition of human scent on the otter's scent-marking site, and avoids confusing your nose with any scents from soap or hand sanitiser. If spraint is removed by hand for sniffing, do not replace it on the spraint site because it will carry your smell. The best thing is to drop it in the river/sea.

2.2.2 Spraint sites

Spraints can be isolated, single deposits apparently left randomly, but most are deposited at spraint sites and these sites are regularly replenished. The spraints at these sites can be scattered about in a small area covering about a square metre or so or may be focused on a particular object that is usually elevated, such as a rock, tree root or tuft of grass. Where there is a particular raised feature, the spraints can be stacked on top of each other creating a small pile. Often, there can be one or more spraint piles and scattered spraints (Figure 2.3). Habitually used spraint sites that are on vegetation tufts can take on a notably brighter green colour than the surrounding vegetation, presumably because the grass has been fertilised by the spraints. The grass at these sites can also appear to have a halo of die-back around the tuft, presumably due to otter urine.

Figure 2.3 Otter spraint site on rocks in foreground, showing several scattered spraints and some spraints forming a small pile.

2.2.3 Latrines

An otter latrine is significantly different from a spraint site. A latrine comprises numerous droppings, which are digestive eliminations as opposed to the token amount of faecal matter in an average spraint. The droppings accumulate to form a significant pile, which attains height and/or spreads horizontally, often covering an area as large as, and sometimes larger than, a dinner plate (Figure 2.4). The form of the droppings themselves can err towards sausages and look much bigger and drier than average spraints (Figure 2.5); these are more likely to be confused with pine marten or maybe mink. The droppings can also be mash in their original form when deposited, or due to washing with urine because otters also prolifically urinate at latrines, or because of rain. The difference in form between a spraint/spraint site and a latrine has only recently been described (Findlay et al. 2023). Previous otter literature often referred to both and may have described spraint sites or spraint piles as latrines.

Latrines are often tucked away under cover (for example within or at the entrance to a holt) and thus are less exposed and can weather slowly. This, in combination with their large size and considerable contents, means that they can have longevity as a field sign even when they are not being replenished. As a latrine breaks down and naturally weathers, the droppings can mesh together to form a solid plate

Figure 2.4 Otter latrine showing continuous mass of droppings with typical lateral spread. The droppings are in 'mash' form, although this may be because this latrine was exposed to rainfall as well as being urinated on. For scale, the camera-trap is 14 cm high.

(Figure 2.6), often incorporating any loose substrate such as disintegrated bark. Over time the black matrix of the droppings gets washed through to leave a considerable spread of bony remains such as fine fish bones and scales. Surveyors need to include an expectation of this field sign during surveys. Weathered otter latrines could be completely overlooked, and active latrines could be mistakenly identified as large badger or pine marten latrines on first glance. This confusion may be exacerbated because mammals may use each other's latrines. For example, we have observed badgers using an otter latrine and vice versa. When a latrine is active it can smell strongly of otter from as far as a metre away. This may well be the origin of the ottery odour that can be detected at some resting places, even if the latrine is hidden from sight. Very old latrines can lose this smell, but if some of the dried matrix is left, the ottery aroma can be reinvigorated with a dousing of water, although this is not encouraged at a resting place because it could deter an otter from using it.

2.2.4 Spraint site or latrine?

A fully formed latrine is easily told apart from a spraint site. However, a latrine does not appear in an instant and so there may be confusion between a developing latrine and a well-used spraint site. Great care needs to be taken to distinguish between the two, as latrines are far more important as a field sign when interpreting survey data (see 3.4.1 Latrines). In freshwater habitats, one or two notably large faecal deposits

Figure 2.5 Drier pellet-type droppings at otter latrines: (a) out in the open and (b) in the shelter of a hollow tree. These latrines have significant mass and lateral spread despite the droppings not being hydrated by rain.

Figure 2.6 Old latrine where otter droppings have adhered to disintegrated bark fragments and have also been enriched by badger dung. When we washed this latrine through a sieve (see insert), the bony parts of the droppings could be seen and the characteristic otter aroma returned.

(Figure 2.7) found in habitat that is suitable for resting, such as dense vegetation or by an enclosed structure, will need more consideration than average spraints. One would question whether it was a chunky spraint or a faecal elimination, and in a survey report this would need to be carefully recorded with a scaled photograph. A repeat visit a couple of weeks later would be required to see if there had been a build-up of droppings in a latrine and whether there was a scattering of more typical spraints or a spraint pile. In coastal habitats, however, spraints can be much chunkier, especially when they contain crab or lobster remains, and a single chunky spraint would not be so noteworthy.

To summarise, the difference between a spraint pile and a latrine is that the former is often on a stone, knobble of root or similar and, while obviously comprising numerous spraints, it does not attain much mass. In contrast, a latrine is larger with lateral spread and is often on flat ground. There can be a focal point but this is usually swamped by the quantity of faecal remains.

2.2.5 Sign heaps

Otters often scrape up small piles of vegetation (Figure 2.8), silt or sand and spraint on top of them (Figure 2.9). These are known as 'sign heaps' and can often be seen

Figure 2.7 Single, notably large faecal deposit of otter together with fragments of other spraints/droppings, which were on an area of overturned common reeds *Phragmites australis*. Nearby, beneath the overturned reeds, was a couch comprising a shallow depression or form in the reeds lined with fresh green growth that was well protected beneath a reed 'roof'.

at a distance due to the scratch marks surrounding them. Sometimes there is no detectable spraint because the scent mark may have been urine. From a single test smell of fresh otter urine it seems that otter urine does not have the same distinctive smell that is so readily detectable in otter spraints. As it is unlikely that a site would only offer a urine-topped sign heap as the sole evidence of otter, not being able to confirm scentless sign heaps is not a problem.

2.2.6 Washing pools and rolling pitches

In coastal habitats, otters have favourite places in which they roll to condition their coats and probably to scent mark from ventral glands on their chins, groins and front feet. These rolling pitches (Figure 2.10) are localised areas of flattened, sometimes dished ground (i.e. a shallow depression), often with the vegetation eroded by frequent use. They are sometimes between tussocks of grass to give the otters more contact as they roll and are often marked with spraint. Often they are close to a small freshwater pool which they rinse their coat in prior to rolling/grooming. They can be found year round.

In freshwater habitats, similar areas of flattened ground can be found, although usually the vegetation remains but has been scratched up and is accompanied by

Figure 2.8 Vegetation scraped up into an otter sign heap with spraint on top. The spraint contained salmon roe *Salmo salar* but the black spraint matrix can readily be seen between the egg pearls. The spraint smelled of otter. For scale the pen is 15 cm long.

Figure 2.9 A sign heap in silty substrate with otter prints and scratch marks.

Figure 2.10 Bare area in vegetation that is a well-used otter rolling pitch in coastal habitat with a freshwater pool behind and otter path leading away.

scattered spraints (Figure 2.11). These can be by a river or along otter paths leading from a river. The need for such grooming areas is less obvious in freshwater habitats and it is likely that they are related more to scent marking. On the River Tweed catchment in Scotland, these rolling pitches abruptly appear in the autumn and look very active for a couple of weeks, before becoming derelict and unrecognisable. Often, they are reinstated in the same location each year.

Freshwater otters also roll in the loose, crumbly bark within old trees, either within the hollow trunk or within wide-diameter, near-horizontal hollow boughs that are either part of the living tree, or large boughs that have fallen to the ground. Otters will often climb trees to a pollard platform to roll, access a large hollow bough or find an opening in the main hollow trunk through which to reach a good rolling spot.

Figure 2.11 A scuffed, scratched-up area of vegetation marked by spraints next to the River Tweed. Best described as a rolling pitch, these tend to appear in autumn and become indistinguishable after two to three weeks. They may be more related to scent marking than coat conditioning.

2.2.7 Footprints

Otters have five toes on each foot resulting in asymmetry to individual footprints. However, footprints are often only partially registered, with an imprint of the fifth toe (or 'thumb') not visible. This needs to be borne in mind when looking at prints with four registered toes. The toes appear oval or tear shaped depending upon the substrate, but there are always spaces between the toes, unlike canine prints where the toes are tight together (Figure 2.12). Illustrations of well-registered otter fore and hind footprints with webbing visible are shown in Figure 2.13, with mink prints for comparison. However, from our experience the webs between the toes don't often leave a distinguishable mark, so the absence of webbing should not be used to rule out otter. The footprints of mink are superficially similar in that they have five toes and are asymmetrical. However, they are often described as star shaped, probably because the central axis of the toes can often be traced towards the 'palm' almost to a common point. In terms of size, a rule of thumb is that an old 50p piece (3 cm diameter) would cover a mink footprint. A comparison of mink and otter prints registered in silt show the size and shape differences in a field context (Figure 2.14).

Figure 2.12 Footprints of otter and dog for comparison. The hind footprint of otter shows five oval to tear-shaped toes, crescent-shaped heel pad and asymmetry of the print. The partial front footprint of otter shows it is slightly smaller than the hind print. The dog print is from an average-sized breed and shows the four toes and a triangular heel pad. The dog print has bilateral symmetry and is generally quite compact.

Figure 2.13 Illustrations of otter fore and hind prints (left) and mink fore and hind prints (right). Scale bar is in centimetres. (David Wege davidwegenature.uk)

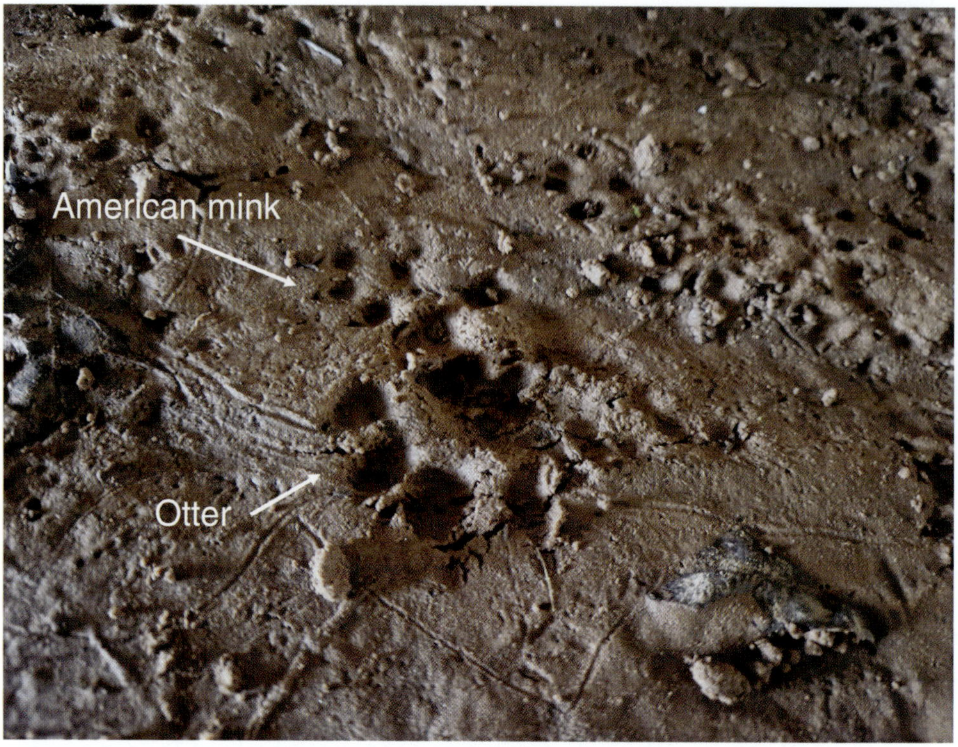

Figure 2.14 Otter and American mink prints registered in silt.

Substrate, age of otter, sex and the otter's gait can affect how a footprint is registered. A study of footprints of nine otters in captivity found that seven measurements coupled with statistical analysis were required to have good confidence of identifying the sex of an otter from hind prints registered in mud (Mercier and Fried 2004). The results of a study of captive otters that measured the forefeet of an adult male, two adult females and a male and female cub at 16 weeks old were counter-intuitive, in that the width of one female's prints exceeded those of the males, and both the length and width of the female cub's prints were similar to those of one of the adult females and larger than those of the male cub (Wayre 1979). Summarising these two captive studies gives an average length of the hind feet (as measured in mud) of the male otters as 8.9 cm and females as 7.3 cm. The length of the male's fore foot in the Wayre (1979) study was 6.4 cm and those of the two females were 4.5 and 5.5 cm respectively. Therefore, the hind feet of an otter have been shown to be larger than the forefeet of that otter. Life-sized illustrations of otter and mink footprints can be found in Wege (2024). More detail about the morphology of Eurasian otter footprints and how to distinguish them from two other otter species with overlapping geographical ranges (Asian small-clawed otter *Aonyx cinereus* and smooth-coated otter *Lutrogale perspicillata*) are given in Kirshner et al. (2022).

Translating these studies into practice indicates that inferring details such as sex and age of the otter that left a given footprint is not usually possible with any degree of certainty. From experience, large and small prints found together can be identified as a female with cub(s), although otter and mink prints together could confuse the enthusiastic surveyor. Repeated encounters of extra-large prints along a reach of river could be *postulated* to be from a large male otter.

The arrangement of prints in a track can indicate the gait/speed of the otter: when walking, the hindfoot is placed behind the forefoot, and at faster gaits the position of the hindfoot moves forward so that it can fall on, partially on, or in front of the forefoot. In soft substrates, including fresh snow, the tail drag can also be seen in addition to the imprint of the short claws, but both of these are relatively rare.

2.2.8 Paths and slides

Otter paths can be difficult to distinguish from badger paths, although there is certainly a difference between the small, tight and often tunnel-like paths exclusive to otters on the west coast of Scotland (Figure 2.15) and the paths used exclusively by badgers, such as those that radiate from a badger sett. It is useful to think of four characteristics when looking at paths: (i) the width of the compressed substrate,

Figure 2.15 Coastal otter path tight and close to the ground, often with squeeze holes through the vegetation.

(ii) the flatness, or otherwise, of the substrate, (iii) the width and height of the vegetation break, and (iv) the position relating to habitats and features.

If the width of the substrate showing signs of contact with feet (i.e. flattened and compacted) is greater than approximately 15 cm, it is likely to be from passes of larger mammals than otter, such as deer, dogs, sheep or people. Larger and heavier mammals such as deer and sheep often leave divots in the substrate, which can also be seen if there is long vegetation on the run. The substrate of otter (and badger) paths tends to be more evenly compacted. The vegetation along otter paths creates a smaller 'tunnel' in cross section in comparison with those along badger paths, as badgers are bulkier animals. This can look as if a log has been lying in the vegetation for some time and then removed, and the vegetation has grown up around it and the ground is compacted. However, the difference between otter and badger paths is not marked and in order to create their own search image a surveyor would need to experience coastal otter paths where badgers are absent. Where the break in vegetation continues above mid-calf height, it is likely to be a larger mammal such as a deer, domestic dog or sheep. Even then, life is rarely that simple, and paths are often used by multiple different species.

Otters often scratch up bits of vegetation to make a sign heap along paths or may spraint at intervals. Rolling pitches can also be found along otter paths, although paths are easier to find in winter and spring than other seasons. Finding these signs is good evidence that the path is used by otter. However, as always, their absence does not mean that it is not an otter path. Finally, the context of a path is important. For example, paths linking a river and a pond, those cutting across a meander loop or rocky headland on the coast, or paths associated with a freshwater pool in coastal habitats are all likely to be made by otter. A path that enters a river and reappears on the opposite bank is less likely to be from otter, and is commonly from deer. A path lacking any additional field evidence of otter such as spraint or footprints should only be attributed to otter by a confident surveyor or where the context shouts otter.

Otters are terrestrial mammals and will often prefer to run beside a watercourse, especially when travelling upstream, against the flow of water. Footprints on habitually used paths can frequently all point upstream (Figure 2.16). Otter paths often form at regularly used shortcuts across meander loops and at regular routes where otters leave the watercourse to reach features such as ponds and marshes. These connecting terrestrial routes do not have to have cover, such as hedges or ditch lines; otters will happily run across an open field or through grassland or crops.

An otter 'slide' is simply erosion of a bank or slope where the otter enters or gets out of the water regularly. It should be remembered that other species also get in and out of the water, especially dogs, and so other corroborating evidence such as spraints or prints is usually needed to be sure it is an otter slide.

2.2.9 Food remains

Small prey items are typically eaten in the river while the otter is foraging. Larger prey can either be eaten in the water or carried to the shore or an island where it is eaten. The remains may later be found by the surveyor (Figure 2.17). Occasionally, larger fish kills can be found some distance from the river – 100 m or more away – and it is difficult to conclude whether an otter dropped the fish or it was taken or scavenged, then dropped, by another mesocarnivore. We observed a case where

Figure 2.16 Otter track next to deep and often fast-flowing section of the River Tweed: (a) otter path that is broader than expected – for scale the pen is 15 cm;

Figure 2.16 (*cont.*) Otter track next to deep and often fast-flowing section of the River Tweed: (b) on closer inspection, all of these otter tracks are heading upstream.

otters would take large fish, caught in a semi-dewatered reservoir, 130 m across land to a tunnel. Camera-trapping showed that badgers were following the otters when they were carrying a fish and a badger *Meles meles* occasionally followed the otter into the tunnel and emerged a few minutes later with the otter's fish. The assumption was that the badgers had been stealing from the otters at the reservoir edge and the tunnel gave the otter a greater chance of keeping its food.

The above descriptions aside, most prey items are small, and the classic fish kill of an otter is not a frequent find in our experience. Such remains of bigger prey items are likely to be scavenged by other mammals and birds and so have limited longevity, although with careful searching, large scales and bones can be found.

The same pattern can be observed in coastal habitats, with otters eating smaller items in the water and often taking larger or difficult prey to dry land. Accumulations of crab, lobster and sea urchin remains are readily found where otters have been feeding, although they can be moved around by gusts of wind. Such feeding remains are often encountered on prominent knowes/knolls – but as this is also where gulls and corvids leave the remains of whelks, cockles and other shells, these should not be interpreted as otter prey remains.

Figure 2.17 Prey remains of otter: (a) fish remains on a pontoon adjacent to a pond; (b) typical coastal crab remains.

2.2.10 Otter hair

It *is* possible to find otter hairs, but they are less abundant, less conspicuous and far more difficult to spot than, for example, badger hairs. Otter hairs are much shorter and finer than those of badger. With a meticulous search, they can be found at rolling pitches and other grooming areas such as those in old willow pollards. Rifling through bedding at a resting place to search for otter hairs to confirm otter presence or resting would require a licence. Overall, otter hair is not recommended as a general field sign, although it may occasionally be found.

2.3 Finding field evidence of otter

Before looking for field signs it is important to consider both recent rainfall and also whether there have been any spates on a river. Heavy rain and spates or other inundations readily wash away spraints and footprints. In the national otter survey of Ireland, rainfall was found to impact detection of otter field signs in running freshwater habitat and had the greatest negative impact if it rained during the seven days prior to a survey (Reid et al. 2013). As rainfall didn't impact coastal or still-water sites, it was suggested that water level rise had more of an effect than the 'washing away' effect of the rain. It is therefore difficult to state a period of time sufficient to allow field signs to build up, as this will vary with the time of year, how the river rises after rain and how the otters are using their home ranges. As little as a few days in busy otter areas may be adequate, but in other areas two or more weeks may be needed. Local knowledge and professional judgement should be used, although in general, surveys should be timed to avoid heavy rain in the seven days prior to survey. Reid et al. (2013) also found that the surveyor's skill had a greater impact than rainfall on whether a site was recorded as positive for otter. Identifying spraints is relatively easy, but there is a greater level of skill required to find spraints; this is where an experienced surveyor will very much out-perform the newbie.

Otter habitat is very variable, including the treeless peatlands of northern Scotland, lowland deciduous woodland, urban landscapes, agricultural landscapes and fenland that is criss-crossed with straight drains and ditches. The general principles of finding otter field signs remain the same, regardless of geographic area and habitat type. When looking for field signs a surveyor needs to think at several different spatial scales. Firstly, the landscape may indicate areas where paths are more likely, such as shortcuts between rivers, across meander bends or terrestrial routes between watercourses, to ponds or to marshes well away from the watercourse.

Next, the surveyor needs to think about large features such as ledges or rocks under bridges, weirs, natural rock falls, spits of land, islands and confluences – all favoured places for otters to spraint. The surveyor then needs to think about the local topography in these areas from an otter's viewpoint, which is about 20 cm above ground. Spraint sites are often on slightly elevated features, such as grassy hummocks, large rocks in a watercourse and on its banks, or even mole hills. In essence, they can be on any features that stand out from an otter's viewpoint, or maybe their nose point! Furthermore, spraint is often perched on the highest point of the feature. The otter's aim is remarkable, and we have observed an otter moving a small pebble and sprainting on top of it with great precision.

Beyond these features that provide attractive sprainting sites, less obvious places are also sprainted on – such as the saddles of leaning trees, ledges in undercut riverbanks, inside hollow logs and by paths. Otters also spraint at features that are relatively hidden from our perspective, such as where a riverbank has been eroded and undercut and the turf curls down forming a section of tunnel, or between the large boulders used for bank reinforcement.

Some habitats, including the fenlands of East Anglia and the Somerset Levels, offer few features such as the weirs, rocks or erosion shelves that are a general 'go to' for spraint checks. However, even along the most managed watercourses that are sided by agriculture or fen habitat, for example reed or sedge beds, there is often an occasional stone, tree or patch of bramble that can yield field signs. Other places to check include the junctions of small ditches with larger Internal Drainage Board ditches and at pipes or outfalls (even small agricultural drainpipes); spraints are often concentrated in gaps in reed or sedge beds, where the vegetation changes and the tall fen species lose their dominance. So, despite the perception of a featureless landscape, spraint can readily be found by an experienced surveyor (Julia Massey, M&H Ecology Ltd, pers. comm. 2025).

Field signs of coastal otters are often patchily distributed, with areas of dense signs in obvious and/or slightly elevated places. There are often concentrations of spraint on small headlands, on the highest knowe/knoll, on small cushions of fine grass turf nestling among the rocks and by freshwater pools. Along the east coast of Britain, where sandy beaches are more common than rocky coastlines, the remains of crab can be found in spraints a few kilometres inland, implying foraging along the shore. Although long, sandy beaches are not optimal otter foraging habitat, runs, spraints and even holts can be found in dunes. Any small rocky features in this type of coastal landscape should be targeted for field signs of otter.

2.4 Covering the ground

A good awareness of the local landscape is essential when commencing an otter survey. The surveyor should use maps, satellite images and good old 'looking around and about' in the field to pick up on features such as ponds, marshes, small ditches, mill streams and mature trees, even if the latter are relatively isolated and in managed grasslands such as parkland. There should be an assumption that all these features are potentially attractive to otters, and they should be investigated whether or not spraint or other field signs are found by the river or shoreline.

2.4.1 Freshwater

The distance that you survey away from your site should follow current guidelines and take into account the potential zone of influence around your development location. Even for a small site, such as a single house, the impact zone should be considered as being 200 m from the site boundary at a minimum. This distance encompasses the risk of the worst-case scenario, which would be proximity to and potential disturbance of a natal or maternal holt.

The most efficient approach when surveying is to have a pair of surveyors working together along one bank, one surveyor covering the water line and riverbank and a second surveyor zig-zagging through the terrestrial habitat. The terrestrial surveyor would examine any areas of structural heterogeneity such as hollow trees, rock piles and dense vegetation, and look for any paths leading away from the river. The surveyors should communicate so that any field signs such as paths found by one surveyor can be followed up by the other. The opposite bank can be undertaken on the return journey. Any paths leading away from the river detected by either surveyor should be followed and investigated.

Upper catchments, whether lowland or upland, can be occupied by otters. Often 'cross-catchment routes' are mentioned in grey literature, suggesting that some watercourses in the upper catchment are used infrequently to cross from one river catchment to another. This is misleading, because it implies that otters do not live in the top of the catchment. Upper catchments, lowland or upland, often offer excellent otter habitat in small streams, ditches, marshes and watercourses and can be continuously occupied. When surveying upland streams for hydro schemes or windfarms, for example, a river may pass through sections of steep-sided, rocky, inaccessible gorge. Although these may be considered natural barriers to migratory fish, otter surveys should continue upstream of the gorge. Other potential prey such as resident brown trout *Salmo trutta* and amphibians are likely to be present in the moorland above. Otters will navigate such gorge sections, and signs of otter and even resting places can be found further up the watercourse. In lowland landscapes, rivers often start as a network of field ditches and/or tiny watercourses or marshes at the top of the watershed. Some of these watercourses may even be seasonal. These areas may be extremely important and may be part of a home range that is used extensively. They may be a breeding area and may provide seasonal food such as amphibians, or foraging habitat when the rivers lower in the catchment are in spate.

It should be noted that a combined otter and water vole survey is rarely adequate because each species requires the surveyor to read the river in a completely different way. Surveys for water voles focus on the area within 2 m of the toe of the bank (Moorhouse 2008, cited in Dean 2021); the surveyor should be completely 'in the zone' for water vole signs and rarely needs to look away from the watercourse. Furthermore, the season for water voles is from mid-April to the end of September (Dean 2021). Most of this period (June to September) is suboptimal for otters because dense vegetation hampers views in terrestrial habitats and can conceal structures, field signs are more difficult to find and otters deposit fewer spraints. Therefore, surveys of these species should not be combined other than in the simplest habitats such as where a river is sided by closely cropped grassland with grazed banks and where any burrows potentially used by otters would be found and flagged for monitoring to confirm use.

2.4.2 Coastal

Along the coast, a source of fresh water is essential for otters to rinse their coats and to drink. This can be a limiting factor in their distribution. Locations of any sources of fresh water can be assessed to a certain degree using maps and satellite images, but otters often use small pools unmarked on maps, and ponded water and freshwater springs under boulders (Figure 2.18) and sometimes even within their resting place. Where water sources are small and ephemeral, the distribution of freshwater can change with the season. A busy otter area in winter furnished with abundant springs and pools can be less suitable in mid-summer when these sources are depleted. It has been noted that otters prefer pools for coat rinsing over running water, such as streams, and also that when there is less freshwater available in summer they spend more time conditioning their coat by grooming and rolling (Grace Yoxon, International Otter Survival Fund, pers. comm. 2025). Equally, subzero temperatures can result in water sources completely freezing, which may affect the seasonal range of otters. Otters can swim under the ice and forage but when watercourses are completely frozen there is likely to be less food available.

As with freshwater habitats, the zone of impact for coastal sites needs to be defined according to the project, and again a good rule of thumb is to include land 200 m from the site boundary. Two surveyors working along transects parallel to the shore works well and is likely to find any paths heading inland. The distance between

Figure 2.18 Strong otter path leading to a small pool of water in coastal woodland habitat. The pool extends beyond view beneath the vegetated boulder.

the transects will depend on the habitat complexity and to a certain degree on how easy or difficult the terrain is. As coastal holts or other important features (such as fresh water sources) can be well inland, surveys should be extended beyond the original notional 200 m, where field signs continue. The intertidal range should be briefly inspected because it can offer excellent substrates for footprint registrations, but searching for spraints among slippery expanses of seaweed and rocks is time-intensive and not very informative, so the surveyor should decide whether it is worthwhile. Any islands and, if possible, fish cage platforms should be surveyed if these would be impacted by the proposal.

Coastal cliffs can be problematic where the tide reaches the base of the cliff or where the sea reaches the cliff base regardless of the tides. The surveyor should focus on searches of the cliff top as otters can ascend and descend steep cliffs. Localities to focus on are where there are freshwater streams flowing over the cliff or in small gullies.

2.5 How useful are field signs to indicate otter activity in site surveys?

Drawing conclusions about the relative abundance of otters from spraint-based surveys may be possible at a landscape scale (Hong et al. 2022) but site surveys should not use spraints to infer otter activity. Using the number or distribution of spraints to draw conclusions on otter numbers, otter activity or the significance/importance of habitat from a typical site survey that is only a small proportion of an otter's home range should not be attempted. This approach has never been validated, and a study on Skye in Scotland found that 'spraint numbers and spraint activity do not correlate with actual otter numbers' (Yoxon and Yoxon 2014). If an area is found in which there is a high density of signs, it is safe to say that an otter has been active in that area. However, if an area has a low density of signs, or no signs, it is not possible to say with any certainty that that area is used less than, or is of less importance, than the area with a large number of signs. This is a very important point, but unfortunately many survey reports write off sections of river as being unimportant because few or no field signs were found, especially small watercourses or ditches and particularly during the summer months. In our view, there is no evidence of a reliably predictable relationship between spraint density and otter activity at a small geographical scale, and drawing such conclusions should be avoided.

There are myriad reasons why spraint is an unreliable proxy for otter activity:

- There is a strong seasonal trend for sprainting, favouring winter and spring.
- Some studies have shown no established relationship between number of spraints and number of otters (e.g. Yoxon and Yoxon 2014).
- There is no relationship between the amount of time an otter forages in an area and the density of spraints (Kruuk et al. 1986).
- Otters have large home ranges and do not scent mark throughout their range all of the time.

- Some habitats are easier to survey than others.
- Some areas have few features to elicit sprainting behaviour while others have abundant features.
- Breeding females may not spraint when using a natal resting place.
- Spraint is affected by rainfall and spate conditions.

This leads to the obvious question: why search for, record and report on spraints during site surveys? There may be a preference from the client or the local authority to record all spraints but ultimately the surveyor should follow recommendations in any survey guidelines when considering whether to invest time recording each and every spraint or just provide examples of field signs from the site survey or any significant field signs such as latrines. However, it is essential that surveyors are familiar with the full range of field signs.

2.6 Key points

- Being able to find and recognise various otter field signs is an important skill.
- Field signs include spraints (faecal scent marks), latrines, sign heaps, washing pools, rolling pitches, footprints and paths.
- Spraints and spraint sites should be distinguished from latrines.
- Otter surveys should be undertaken in good conditions. The river should not have been in spate and there should not have been heavy rain prior to the survey because both can wash away field signs. The surveyor's knowledge of the area should be used to gauge the period of time prior to the survey when good conditions are needed.
- Otter surveys should not be undertaken at the same time as water vole surveys, because the spatial scale of the two surveys is very different.
- Field signs are a way of confirming the presence of otter. However, the density of spraints should not be used to speculate on whether some areas are better habitat than others, used more by otters or that there are more/fewer otters.

3. Resting places and breeding places

3.1 Introduction

Otter holts and couches are often recorded in otter surveys, holts being a hole in the bank or similar, marked by spraints, and couches being flattened areas of vegetation also marked with spraints. Both these terms are well known and well used and are essentially descriptors of field signs. When the Habitats Regulations came into force in 1994 there was a need to identify 'resting places' of otter. This term is applied to the strict protection of a suite of species and so there are no species-specific definitions. Thus, holts and couches were reasonably assumed to represent resting places. Without a method of direct observation to test the assumption that these terms are equivalent, we would be none the wiser. A holt or a couch would be considered a resting place and spraint at a structure would be strongly indicative of a resting place.

Occasionally, during a survey, field evidence comes together perfectly so that conclusions can readily be made (see Figure 3.1), but this is relatively rare. Early radio tracking studies indicated that otters frequently changed resting place, often resting for one or maybe two days before moving to another resting place, while often the actual place of rest could not be located using field signs (Green et al. 1984). This suggests that there is a high level of uncertainty around the feasibility of a surveyor being able to find and identify resting places from a walk-over type survey. The uptake of camera-traps in the twenty-first century has enabled direct observation of holts and couches, giving an accessible way to study and increase the understanding of how otters use them.

Terminology is discussed in section 1.4, but as a quick recap we use descriptors for the type of structure that it is (holt or couch) in combination with the function (natal, maternal, resting place) if known, and use a prefix of 'potential' where the actual function is unknown. Determining whether a structure is used as a natal and/or maternal resting place is not usually possible from field surveys except on very rare occasions, for instance if a dead cub is found outside a structure or if young otters can be heard within the structure. Thus, a field walk-over is very limited when it comes to identifying natal and maternal structures with confidence and the prefix 'potential' is still needed.

Figure 3.1 A suite of field signs: (a) otter tracks along a sandy beach; (b) the drag marks made by an otter carrying either bedding material or a large food item; (c) perfectly registered otter footprints; (d) fresh spraint; (e) signs lead to a burrow entrance in a sand dune; and (f) burrow entrance is round and has spraint outside, and the scuffled sand indicates recent presence, likely to be otter. (Steven Parker and Lisa Thomson, Nàdarra Ecology Ltd).

In this chapter, we draw upon our research programme (2015–2020), which used camera-traps and field sign data at sites to test whether there is a consistent field sign signature that could be used to identify resting places. In other words, whether potential holts and potential couches are always used by otters for resting, and whether all resting places present field evidence as per a holt or couch.

3.2 Legislation and official guidance

We touched briefly on legislation in Chapter 1, but here we present the UK legislation and ecology of the otter side by side to see how best we can interpret the findings from walk-over surveys in the context of the legislation.

Otters are protected by either one or two pieces of legislation in the UK, according to the constituent country. In England and Wales, the otter is still listed on Schedule 5 of the Wildlife and Countryside Act (1981), which refers to protection measures for each 'structure or place which it uses for shelter or protection'. In Scotland and Northern Ireland, the otter has been removed from Schedule 5 and all protection is via the Conservation (Natural Habitats, etc.) Regulations 1994 (as amended).

Legislation uses the term 'resting place' adopted by the EU, or the phrase 'structure or place which it uses for shelter or protection' adopted by the Wildlife and Countryside Act (1981), as amended, to protect the places where otters sleep/rest. You would therefore hope that there would be clear guidance on how such important sites could be readily found and identified, so that all those involved on the ground could confidently go forth and protect such structures according to the law. However, there is no otter-focused guidance in the UK to accompany either the 1981 Act or domestic legislation shaped by the Habitats Directive to guide us in what resting or breeding sites actually are or how to identify them. It therefore seems sensible to turn to guidance on the strict protection of Annex IV species of the Habitats Directive EU (European Commission 2021, 2007) and take what we can from its definition of resting places. It should be remembered, though, that the guidance is not otter specific; rather it explains the underlying rationale of the legislation and how it can be applied across a range of species with very different ecologies (e.g. great-created newt *Triturus cristatus* and otter) and gives examples of how it can be interpreted.

3.2.1 Resting places

Relevant points from the EU guidance on the 'strict protection' of otters (European Commission 2021, 2007) is summarised as follows:

- Resting places must be clearly known.
- Resting places are used repeatedly, even if not occupied all of the time.
- They are sites that the species uses when not active; for otters, this means when they are resting or sleeping.

While we could get stuck in an argument regarding otter resting places being synonymous with the common perception of a 'holt', which is specified in the guidance, using field evidence to identify holts as otter resting places is clearly too simplistic. However, the guidance heavily implies that a resting place is where an animal sleeps, rather than a favourite spot that an otter may use to loaf around in while awake without exhibiting more active behaviour such as foraging, sprainting and grooming.

3.2.2 Breeding sites and places

In the UK, the term 'breeding site' is used (Liles 2003), and at first this may seem more intuitive than a resting place because it must involve a breeding female with cubs. The context of that publication (otter breeding sites, conservation and management) was to conserve *breeding sites*, which are defined as undisturbed blocks of habitat that include one (or more) potential natal dens and access to a good food supply, together with a lack of flooding and play areas for cubs. However, for otters (maybe more than other species) the concept of a breeding site is more complicated when trying to apply it to the protection of any such sites as part of ecological impact assessments or similar.

We make a pragmatic distinction between a 'breeding site' (as defined by Liles 2003) and the collection of resting places the female uses for giving birth and rearing her young cubs. This distinction does not devalue the importance of the supporting habitat; instead it recognises the difficulty in identifying this habitat. We use the terms 'natal resting place' and 'maternal resting place' to give emphasis to resting places with elevated importance, and refer to these collectively as 'breeding places'. Breeding sites, natal resting places and maternal resting places are discussed below.

There are two key questions that need to be considered when defining a breeding *site*:

1. Young otters stay with the female for approximately a year before they become fully independent, so should all sites used for resting by this family group while the cubs are dependent on their mother be treated as breeding sites?
2. What is the spatial extent of the breeding site? Should it be considered as the few square metres that the female and cubs rest on, or should it also include surrounding habitat that protects the place(s) of rest and, if so, how far should that extend?

When addressing the first question relating to the period of time crucial to the survival of a family group it goes without saying, and is clearly stated in the EU guidance, that this includes the actual place of birth (European Commission 2021, p. 42), often called the 'natal holt'. To ensure successful breeding, females with young cubs that have emerged from the natal holt need to be considered as vulnerable and require increased protection. Impacts on a structure that houses a female with cubs of, say, three months old that are barely out of the natal holt almost certainly have higher risk of cub mortality than if the cubs are ten months old and spending time foraging away from the female.

When defining the time period during which sites used for resting by an otter family might be considered to be breeding sites and therefore that they merit more extensive protection due to the otter family's vulnerability, it is useful to reflect on what we know about cub development. Otters start to wean on to solid food around the time they first emerge from the natal holt – at approximately eight to ten weeks old. In captivity, weaning takes approximately two weeks (Heap et al. 2010), but there is an absence of information on the weaning period in wild otters in published literature. We have observed a female with suckling marks ten weeks after her cub had emerged from the natal holt. This suggests that the cubs are not fully weaned in the wild until they are at least five months old, assuming the cub first emerged from the natal holt at about ten weeks old.

After the cubs have left the natal holt but are still very young, the female will leave them at resting places for long periods while she forages without them. Protection measures need to encompass these early days while the cubs remain reliant on the female. We suggest that families with cubs younger than six months old may find it difficult to reunite if separated by disturbance. Resting places used by females with cubs aged up to approximately six months should therefore be considered as breeding sites but may best be described as 'maternal resting places' to separate them from natal resting places. While this may seem like a long time, it has some parity with the protection of breeding European badgers and their setts in the UK, where birth is estimated to peak in early February and with the cubs staying underground for approximately eight weeks (Clark 1992; Dugdale et al. 2008). In the UK there is a presumption against works that could disturb breeding badgers during pregnancy, birth and early cub development, which is between December and June and which covers the period up to when cubs are approximately five months old.

When thinking about the second question relating to the area that should be considered as a protected breeding site, natal and maternal resting places should be distinguished from each other. This is because natal holts are likely to be a rarer resource in a female's home range due to the specific characteristics they require, and there is commonly just one holt in use during any breeding cycle, unless the female is disturbed and moves the cub to a different structure. One of these characteristics is that they need to be undisturbed. Fortunately, NatureScot has offered some guidance on this. At the time of writing, a protective exclusion zone of at least 200 m is required during development, so that breeding sites (i.e. natal and maternal holts) can be left undisturbed (NatureScot 2024).

As the female moves her young family after leaving the natal structure, the range of the female with her cubs increases and the number of resting places used increases to facilitate this. These resting places are more common than natal structures, and it can be postulated that the characteristics of these places are less exacting than those of a natal structure. But this is where it gets tricky. If a female is ranging with her four-month-old cubs and has a number of resting places along 5 km of watercourse, do we consider this entire 5 km stretch to be a breeding site? This can't be applied because we cannot demonstrate her range without radio tracking. Instead a more practical approach is used that requires a third question: why do we need to know? If we have a project that is spatially limited and short lasting, such as repairing a section of riverbank, breeding sites should be considered spatially (i.e. are there any resting places within the project area and 200 m from its boundary?), but also temporally (i.e. is there a young otter family occupying this reach of river at the time of works?).

If you find a potential resting place you can avoid breeding otters by ensuring good timing, using camera-traps to identify use of the resting place by otter families with small cubs, and by programming works (covered by a licence) to start when the cubs are of a robust age – approximately six months old. Therefore, it is essential to estimate age from camera-trap footage (see section 6.5.3). Larger construction sites, longer periods of construction or projects that will permanently affect habitats or increase footfall (such as housing schemes near rivers or standing water, or even new footpaths) need more careful consideration. The information baseline, which informs the mitigation, will need to include much more than a couple of walk-over surveys to demonstrate whether there will be any impact on breeding, so that the temporal and spatial variation in breeding is properly covered.

3.3 Expectations of otter field surveys

There has been a general expectation that otter surveyors can make one or two field visits to a site and use field signs to identify all resting and breeding places. This expectation, combined with the long-standing reliance on field signs, has, in our view, led to over-interpretation of field signs and the over-expectation that a field surveyor can easily find and identify all otter resting places. It is time that expectations of what field ecologists can and can't do become better aligned with published studies. Two complementary types of study have improved our understanding of resting place use: radio tracking, which gives insight into how individual otters use resting sites across their home range, and camera-trapping, which helps us understand in depth how individual resting sites are used by one or several otters.

There is general parity across radio tracking studies that follow resting place use by individual otters. A key pattern is that some resting places are used far more frequently than others, and that otters rarely use the same resting place on consecutive days (Green et al. 1984; Rosoux and Libois 1996). Green et al. (1984) also found that a third of resting places accounted for two-thirds of recorded rests and approximately one-third of resting places had only a single rest during the radio tracking sampling period, which varied per individual otter from 22 to 98 days. Furthermore, studies on Shetland found that some holts were favoured for resting over others and even suggested the categorisation of holts as 'main' (used regularly), 'intermediate', and 'subsidiary' (infrequent use) (Moorhouse 1988).

In our research we camera-trapped 26 potential resting places across the River Tweed catchment for approximately a year at each site and regularly and systematically monitored any field signs around each (Findlay et al. 2023). Field signs were used initially to find these 'potential' holts or couches; each had a discernible area suitable for resting and spraint was present within 1–2 m.

Several of our monitored structures looked like textbook holts – an enclosed structure that had strong field evidence such as spraints and otter paths associated with them. When we used camera-trapping observations to record the amount of time each otter stayed in the structure we found they were visited by otters very frequently, as the field signs suggested, but that the otter usually stayed less than a minute at the site.

Rests (i.e. spending a substantial period within the structure) were never observed, so these structures that had all the hallmarks of a holt were either used solely as spraint sites or had another function that was not obvious. Figure 3.2 shows two of these 'non-resting' structures with an account of their field evidence and number of nights otters were recorded.

The confirmed resting places we identified from camera-trap data in the River Tweed catchment were all used repeatedly (Figure 3.3). There was variability in the frequency of resting, but all structures were used more between winter and spring. This may be because all these structures were enclosed or semi-enclosed and therefore offered shelter when temperatures were lower. We also monitored a holt in Fife, lowland Scotland, which was occupied remarkably often, including repeated use as a natal and maternal site, and by non-breeding individuals between 2009 and 2025 (it is still being monitored, but several years' use is reported in Findlay et al. 2017).

This radio-tracking and camera-trapping research helps to shape expectations of what a field ecologist can realistically achieve when undertaking a site survey, which is often just one or two visits to a small proportion of an otter's home range. For starters, it is unlikely that surveyors will find, let alone identify, the third of resting places that are only used once. However, it could be argued that these are

Field signs: path and spraint (at entrance and or on stones by water) at every 3-weekly survey visit.

From camera-trapping:

- Presence recorded on 92% camera-trap nights (n=373)
- Occasional family groups
- Many visits each night, up to a maximum of 8
- Max time in structure was 3min
- No rests recorded

Field signs: path and spraint (at entrance and or on root by water) at every 3-weekly survey visit.

From camera-trapping:

- Presence recorded on 38% camera-trap nights (n=340)
- Occasionally multiple visits each night, up to a maximum of 4
- Max time in structure was 1min
- No rests recorded
- Badger occupied for 13% camera-trap days

Figure 3.2 Structures that suggest resting places but which were not used for resting: (a) otter recorded over 90% of camera-trap days and strong field signs of otter always present, and (b) otters were recorded frequently going in and out of this large, hollow ash tree and there were always abundant field signs of otter. However, otters never used it for resting during a year of monitoring, although a female badger rested there daily for a few weeks.

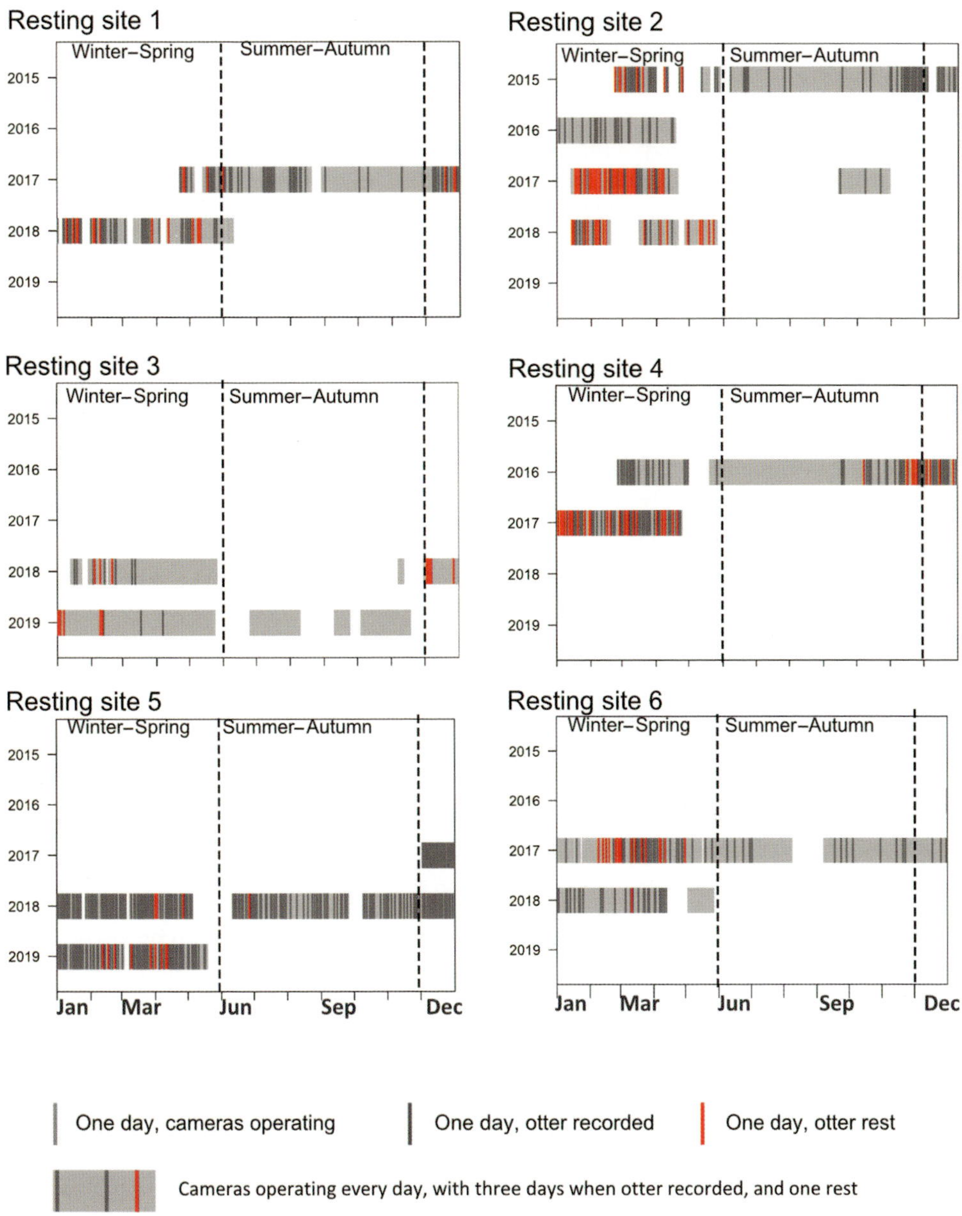

Figure 3.3 Colour-coded calendars of six resting places on the River Tweed catchment, illustrating resting and visiting of study sites by otter. Use for resting is almost exclusive to the winter–spring period, even though otters also visited these structures at other times of the year. Taken from Findlay et al. (2023) (CC BY 3.0).

not protected by legislation, as there would be no expectation of future rests. Finding the sites that make up the third of resting places used habitually, which potentially comprise two-thirds of rests, is a far more realistic aim for surveyors. The resting places we camera-trapped are likely to be among the resting places used habitually and are arguably more significant. As for the theoretical third of resting places that remain, ecologists are likely only to find these by luck.

3.4 Identifying resting places during walk-over surveys

The following is largely based on analyses of the field signs found at our 26 sites monitored by camera-traps. Analyses of otter activity at these sites found that seven were resting places, one was a possible resting place and 18 were essentially spraint sites rather than resting places (Findlay et al. 2023).

3.4.1 Latrines

If a latrine is encountered during a walk-over search for resting places, it is almost certain that you have found a resting place and therefore it is protected by law. In our research, we noted that otters used latrines only in association with resting, for defecation and urination prior to, during or after a rest. If an otter visits a site without resting, any spraint is usually deposited in a different area and not on the latrine. The surface of any latrine should be examined very briefly to judge whether the latrine has been recently refreshed, which would indicate that the resting place has been used recently or is in current use. There could be individual wet droppings on the surface, which would be evidence of very recent use (Figure 3.4a); often, the individual droppings are all smooshed together due to rainfall, otter footfall or otter urine (Figure 3.4b). Old latrines that haven't been refreshed can comprise a mass of pale food remains in which the matrix of the droppings has been washed out. A refreshed latrine is evidence of recent resting place use, but if the latrine contents look old it should be remembered that there could be other hidden latrines, for example inside the structure itself, with recent droppings on. Also, where you do not see evidence of recent droppings on a latrine, there should still be an assumption that the resting place will be used again and that the structure is protected by law, although it could be months hence before it is used for resting.

Not all the resting places we identified had latrines that could be seen by surveyors, because some were concealed within the structure; latrines within holts have also been recorded on Shetland (Moorhouse 1988). We also found that if a structure is very close to water, a latrine area could be washed clean by spates and thus only recognised some of the time. Therefore, if the entirety of a potential resting place cannot be properly examined, or if the entrance area of the structure is washed over regularly, a structure cannot be discounted as a resting place if a latrine is not found. Such structures should be treated as potential holts. There is further discussion of how to deal with such scenarios in Chapter 6.

3.4.2 Bedding and nests

Otters can build sizable nests in their sleeping chambers. Accumulations of bedding approximately 50–60 cm in diameter can be expected (Figure 3.5), although the bedding is often shaped to fit the chamber it is in. However, badgers can build similar nests away from their setts, either within hollow tree trunks or boughs or on the ground where there is cover from scrub or a fallen tree (Figure 3.6). These are virtually indistinguishable from otter nests. To complicate matters further, some structures are used at different times by both species, and bedding can be collected by one or other, or both. Badgers are more efficient at collecting bedding and nest-making and can

Figure 3.4 Otter latrines: (a) latrine tucked away with fresh droppings indicating recent use, and (b) disintegrated droppings (note the large size, approximately 70 cm across) with no fresh droppings. Further latrine images can be seen in Chapter 2.

Figure 3.5 Typical otter nests: (a) within hollow tree where badgers had also taken in bedding, and (b) under rocky overhang.

Figure 3.6 Other nests: (a) badger 'day nest' of approximately 1-metre diameter, although less extensive versions are likely, and (b) nest made of greater woodrush *Luzula sylvatica* on a ledge near a river likely to be otter and/or badger

take in a good amount of material in a short space of time. This can then be enjoyed by otters at a later date. In such cases we did not always observe otters adding to the bedding chamber – they simply used the badger's nest. Thus, the presence of badger hairs at a site, even in the bedding material, must not rule out the potential for that nest being used as a resting place by otters.

When badgers collect bedding material they often drop some of it as they drag it over the ground, where it tends to clump together appearing as a loose accumulation of material. This is rare with otters, because most bedding is carried in the animal's mouth, although the backwards 'badger method' has been observed in otter. We have also observed an otter trying to pull in a partially buried item of clothing by hauling it backwards. Otters tend to collect bedding very close to the holt. Occasionally they pull at vegetation around the holt entrance and so the composition of the nest tends to reflect that. Seaweed is commonly taken into coastal holts and on Shetland, plastic fertiliser bags have been recorded as being used as bedding.

While watching different otters collect bedding over the years at our long-term study site we noted that some used a different route in and out of the holt when collecting bedding to their usual route to and from the river. Presumably, the vegetation behind the holt was preferred as bedding. Chewed-off vegetation around a potential holt is, however, not a good sign of otter occupation unless the surveyor is very familiar with the particular holt. Even then, herbivores cannot be ruled out. The origin of loose bits of bedding/vegetation near a structure is therefore difficult to pin down and could have been dropped by badger or mink, or could even be flood debris that has been blown about.

It should be noted that a nest at a structure is not necessarily indicative of breeding, as all otters may collect bedding. We have even observed captive males attempt to collect bedding even though they were supplied with a cosy, heated holt with ample bedding inside. They would climb shrubs and snap off little branches to take into their holt, indicating that the bedding collection urge is innate for both males and females. A nest in a holt is, however, a strong indication of a significant structure, in that it is likely to be used regularly. For more on bedding collection, see section 6.6.1.

3.5 Identifying breeding places during walk-over surveys

3.5.1 Natal resting places

Finding and identifying a natal holt is the holy grail of otter surveying because these holts are so vulnerable yet very challenging to locate. The jury is split when it comes to field evidence at natal holts. There is some evidence to suggest that the female does not spraint outside her natal holt when she is using it (Durbin 1996; Green et al. 1984), and this is our experience of two natal holts. Furthermore, Green et al. (1984) found that spraint counts within 100 m of a functioning natal holt were far lower when cubs were confined underground compared with the periods before their birth and after their emergence. However, Taylor and Kruuk (1990) found a considerable pile of spraints within 1 m of a site thought to be a natal holt in a reedbed. Although our observations indicate that spraint is a rarity at natal holts, we did observe a male otter scent marking (spraint, urine and rolling) at the natal holt we have monitored since 2009 (unpublished data).

An otter nest would be expected at a natal resting place but its presence does not distinguish a natal resting place from a maternal resting place or even a regularly occupied 'casual' resting place used by non-breeding individuals. It is also likely that most natal holts cannot be inspected in full and therefore a nest may not always be visible.

Natal resting places in relation to the landscape are reputed to be well away from water in both coastal and freshwater systems. The natal holts we have found, which we knew over many years, never flooded, even after the most severe rain events. It makes sense, therefore, when considering the potential for a structure to be a natal resting place, to consider the safety of the pre-emerged cubs. While females have been known to move cubs between holts prior to emergence (Rosoux and Libois 1996; Taylor and Kruuk 1990), we have not observed this. Slow-moving water, good terrestrial habitat, low human footfall and lack of flooding suggest good natal habitat. However, as so few natal holts have been described in scientific literature it is difficult to draw strong conclusions and there will always be exceptions to any generalisations. It is also worth remembering that a female does not need pristine habitat to breed. Breeding is a biological imperative and she will therefore do the best she can when choosing her natal holt.

3.5.2 Maternal resting places

It is not possible to tell from a latrine whether a resting place is occupied by one otter or more than one, such as a family. There is no evidence to suggest that young otters produce differently sized *droppings* from adults, although received wisdom has it that females and cubs produce larger *spraints* than adult males. This is thought to be because males must eke out their reserves of spraint to mark resources in their range more frequently than females. However, we now know that droppings and spraints are different entities (see section 2.2.3), so the size of droppings at latrines is not informative.

As both males and females collect bedding, otter nests are likely at breeding places due to an assumed increased need for thermal insulation for the cubs. They are also present at some resting places not used by females with cubs. The only way to ascertain if a site is a breeding location is to directly observe activity patterns or occupants, which is readily achieved by using camera-traps, as discussed in chapters 4–6.

3.6 Field characteristics of resting place structures

3.6.1 Entrance

The entrance to an enclosed resting place is not necessarily a tunnel but is often a pinch point or 'squeeze hole', for instance a narrow opening allowing access into a hollow tree. The smallest pinch point we have measured was approximately 10 cm across and was only seen to be used by a female (Figure 3.7). These pinch points are rarely circular, but the otter will orientate itself according to the shape of the entrance, and we have observed a large dog otter squeezing itself sideways into the tall, thin entrance of a hollow tree holt. Tunnels need to be somewhat larger, because otters need to be able to move along them. In a study in Shetland, where

Figure 3.7 Entrance to resting place within the trunk of a hollow crack willow *Salix fragilis*, used by a female otter. The width of the squeeze hole is approximately 10 cm, the metal rule in image is 15cm. To avoid disturbance the photo was taken when the resting place had not been used for several weeks, hence the leaf litter.

otter holts originated in fractures in peat, holt entrances were wider than their height, averaging 27 cm wide and 17 cm high, and had a minimum width of 11 cm and minimum height of 10 cm (n = 53) (Moorhouse 1988). As a good rule of thumb, structures with holes where any dimension is less than 10 cm should be discounted as resting places.

Repeated passage of wet otters into a structure and dry otters out, particularly if the otter has to squeeze through the opening, can often smooth the entrance walls of a resting place if in earth or mud. The area outside the entrance can appear compacted and the base of the tunnel can be flattened, again depending on the substrate. If the otter has to climb to an entrance hole up a tree, a muddy trail may be seen, which is even more obvious in snow. Moorhouse (1988) noted that the entrances of active holts in peat were smoothed. However, as this just indicates regular passes of otter, additional evidence would be needed to categorise the structure as a resting or potential resting place. A burrow or structure entrance with any polishing should always be followed up.

3.6.2 Sleeping area

The sleeping area needs to be large enough to accommodate an adult otter, which is around 1 m long, nose to tail tip. Otters tend to sleep curled up in a ball, although we have also observed them sleeping stretched out on their fronts and on their backs. A minimum size of sleeping area is approximately 50 cm diameter although the sleeping chamber can be much larger and is not necessarily a regular shape. Eight sleeping chambers within holts on Shetland were found to be longer than they were wide, with little or no effort made by the otters to modify the shape (Moorhouse 1988). Most of these chambers contained bedding. It would be reasonable to expect that a sleeping area occupied by a female and cubs, especially when they are well grown, has to be bigger.

3.6.3 Disturbance

There are no set rules when considering what may be 'too much' disturbance for an otter to use a nearby structure for resting. Disturbance by people and dogs can certainly be an issue and we have recorded on camera-trap a female and grown cubs abruptly vacating a quiet, rural resting place when a dog walker passed approximately 25 m from the holt. However, urban resting places are common and even natal and breeding places occur in cities. It would seem likely that otters become habituated to a level of activity in their home range and so a stimulus that may not cause any reaction in an urban habitat may elicit a negative response in a quiet rural location. Thus, the potential of any site should not be dismissed due to perceived disturbance, unless that disturbance is extreme. It should also be noted that otters are not active throughout the whole night and will have short nocturnal rests. It is likely that they will use features at night, at a time when there is no human footfall, which they would otherwise not use during the day. Radio tracking research in Austria found that resting places were within stretches of river with a 'wider vegetation belt' (the width of natural or semi-natural vegetation measured from the waterside) when in areas with greater disturbance (Weinberger et al. 2019), although that study did not include analysis of nocturnal resting places.

3.6.4 Surrounding habitat

It is best to keep an open mind when judging whether the vegetation surrounding a structure is 'suitable' for resting or breeding. Two of the resting places we located were surrounded by grazed pasture with a short sward, while others were in rank vegetation (nettles), woodland, a row of riparian trees, coniferous plantation and scrub. Green et al. (1984) found that sallow scrub was a favoured habitat for couches, and Kruuk et al. (1998) found that islands accounted for the highest number of rest days at couches – 37% of all rest days recorded (n = 402). Other habitats used for resting included reedbeds, coarse vegetation (rushes *Juncus* sp. and tufted hair-grass *Deschampsia cespitosa*), shrubs, bank ledges, fallen trees, under boulders and a derelict car. Resting places in peatlands are often surrounded by heath and/or bog with no tree cover.

Contrasting again, there is a holt in a Scottish city in a pile of rubble, put there to provide protection from riverbank erosion. It is in an open area with no tree cover and in full view of passers-by and the otters busy themselves in the river while onlookers chat nearby. The best approach is not to pre-judge the suitability of habitat. Habitat assessments are not recommended as predictors of otter presence, resting or breeding, and there is no published literature that adequately covers what is and isn't good terrestrial otter habitat. It is easy to look at a reach of rural river and say it is suitable, or ideal, or optimal, but dismissing habitat or labelling it suboptimal is completely subjective. We have camera-trapped otters using concrete mill channels with negligible riparian habitat and, often, negligible water (Figure 3.8). These channels frequently ducked beneath the urban surface in long, rubbly tunnels where there were no people or dogs, enabling the otters to commute beneath the townscape between sections of the urban main river. They were able to rest in complete safety during the day, a short stone's throw from McDonald's and Tesco.

3.6.5 Resting place gallery

Sites that were used for resting or breeding that we monitored with camera-traps in our main River Tweed catchment study are shown in Figures 3.9 to 3.14. These match the sites labelled sites 1–6 in Figure 3.3 showing patterns of their use.

Additional resting places from a wider geographical area, identified either from camera-trapping or from the presence of bedding and/or a latrine, are shown in Figures 3.15 to 3.21.

The figures in this section illustrate the variety of structures that can be used by otter. It should be noted that only one of these is an open structure (Figure 3.19) and that resting was only recorded once there; the otter field signs had accumulated because it was mainly used for grooming, feeding and scent marking. The absence of a nest or latrine would be due to its very infrequent use. While we present resting places as a black-and-white scenario, as with most field ecology there will always be grey areas.

Figure 3.8 Sections of an urban mill lade (channel) with minimal riparian habitat, both frequently used by otter: (a) a frequently used otter travelling route leading to underground sections likely to offer important refuges in an urban landscape, and (b) underground sections of the mill lade, an ideal refuge beneath a busy town.

Figure 3.9 An otter resting place confirmed by camera-trapping, annotated to show location of field signs. Spraints were present on every field check and the latrine was present on 11 out of 15 field visits. It was used as a maternal holt, by one or more adult males, and by non-breeding individuals. Its pattern of use can be seen in Figure 3.3, where this is labelled 'Resting site 1'.

Figure 3.10 An otter resting place confirmed by camera-trapping. A latrine was inside the structure and was not visible without an invasive search. Occasionally, nesting material could be seen. Spraints were recorded on 66% of field visits. It was used as a maternal holt, by one or more adult males and non-breeding individuals, and also rested in by mink and badger. Its pattern of use can be seen in Figure 3.3 where this is labelled 'Resting site 2'.

Figure 3.11 An otter resting place confirmed by camera-trapping: (a) wider view of old willow tree, and (b) close-up view of entrance about 1 m above ground. Spraints were rarely found. A latrine and bedding material were visible only through an invasive search, but we only observed a badger taking bedding in. Not used by otter family groups. Its pattern of use can be seen in Figure 3.3 where this is labelled 'Resting site 3'.

Figure 3.12 An otter resting place confirmed by camera-trapping: (a) wider view of old alder tree, and (b) close-up view showing location of camera-trap emplacement facing one entrance at ground level. There was a second entrance about one metre up the tree. Latrines were present adjacent to the nest and at both entrances, and spraint was always present. It was used as a maternal holt, and by non-breeding individuals. Its pattern of use can be seen in Figure 3.3 where this is labelled 'Resting site 4'.

Figure 3.13 An otter resting place confirmed by camera-trapping. A latrine was present in the entrance to the main hollow, and the sleeping chamber entrance was at the rear of the trunk, with the otters having to climb slightly. Spraint was always present, but the latrine was recorded on 20% of visits. It was used by two adult-sized otters as well as single otters. Its pattern of use can be seen in Figure 3.3 where this is labelled 'Resting site 5'.

Figure 3.14 An otter resting place confirmed by camera-trapping. Spraint was present all the time but often just one or two. A latrine was present inside the structure, revealed when the tree was blown over. It was used as a maternal holt, by one or more adult males and by non-breeding individuals. Camouflage camera-traps were designed to reduce visibility from a nearby road. Its pattern of use can be seen in Figure 3.3 where this is labelled 'Resting site 6'.

Figure 3.15 A confirmed otter resting place within rocks in freshwater habitat: (a) entrance, and (b) a semi-hidden latrine visible between rocks.

Figure 3.16 A confirmed otter resting place under a boulder in coastal habitat, with latrine present outside.

Figure 3.17 Location of a confirmed otter resting place within an artificial boulder pile in an urban freshwater habitat. This was confirmed through direct observation and was used as both a natal and maternal holt and by non-breeding otter(s).

Figure 3.18 A confirmed otter resting place beneath a large pile of forestry brash in freshwater habitat, marked by a latrine.

Figure 3.19 A confirmed open resting place on a level platform by a river, with one rest of 17 min recorded: (a) site during daylight with hiking pole for scale;

Figure 3.19 (*cont.*) (b) image from camera-trapping showing otter sleeping at the location (second image: Sara Toule and Anna Redpath, SLR Consulting Ltd).

Figure 3.20 Camera-trap image of otter emerging with dry, fluffy coat from a confirmed resting place in a disused badger set, where it had rested for four hours.

3.7 Finding potential resting places during fieldwork

Given what we know about the low likelihood of finding resting places used for one or two days, and with otters having no fidelity to them, the surveyor (clients and the competent authorities) needs to accept that only the proportion of resting places with some level of habitual use can reliably be found. Green et al. (1984) found that 95% of all resting places were within 50 m of the riverbank. This is echoed by

Figure 3.21 A resting place in a reedbed where the otters have 'excavated' a sleeping chamber within reeds *Phragmites australis* that were cut and discarded as part of the reedbed management (Julia Massey, M&H Ecology Ltd). This type of structure is sometimes referred to informally as an 'otter igloo'.

an Austrian radio tracking study that found that 95% of resting places were in the riparian corridor. As a guide, freshwater surveys should focus more effort on the 50 m of adjacent habitat unless ottery features such as oxbow lakes or ponds are beyond this 50 m. The first stage of fieldwork should identify potential resting places and so particular features should be sought that offer accommodation that is dry and large enough. This search should include:

- Inspections of all large trees, checking for entrances into hollow trunks and boughs that can be at ground level or further up, such as at a pollard platform. It is likely that otters will habitually use the same route to climb the tree and this may become abraded and/or muddy.
- Inspections of large lime trees with epicormic growth (Figure 3.22). We have found spraint-marked runs into epicormic growth with well-used 'tunnels' climbing upwards and inwards. We have camera-trapped otters entering the mass of twigs and flood debris that can get caught on the dense growth of such trees. According to keepers, an otter in captivity once chose to have her cubs in a large lime tree and took copious amounts of provided bedding material into the tree, although it was not clear if the nest was in the twiggy growth or whether there was access to a hollow trunk.
- Eroded banks with ledges and/or burrows, often behind tree root systems.

Figure 3.22 Mature lime tree: (a) substantial epicormic growth of lime trees provide refuge and resting potential, or may conceal entrances into a hollow trunk, and (b) otter paths, marked with spraint, can be found leading into the mass of epicormic growth and can be followed as the path climbs upwards forming a tunnel in the dense mass of twigs and debris. Note that similar signs of mink can also be found, and that trees may be used by both species.

- Islands.
- Brash piles from forestry work.
- Piles of flood debris, often caught up by a bankside tree (Figure 3.23a).
- Under garden buildings such as sheds and decking.
- Disused badger setts.
- Rock or boulder piles, both natural and human-made.
- Under single boulders.
- Burrows and concealed ledges associated with bridges (Figure 3.23b).
- Areas of dense scrub such as bramble and gorse.
- Areas of dense seasonal vegetation such as purple moor-grass, bracken, tall ruderals.

Some features can be fully walked around, assessed and discounted, such as a large ash tree with no rot or other holes providing access into a hollow. However, other structures are more difficult, such as areas of dense gorse scrub or a large pile of flood debris. In these cases, the surveyor needs to find evidence of otters going into the structure or habitat, such as a defined squeeze hole or run. With difficult scenarios such as this, a squeeze hole or run is not sufficient to state that the structure is a resting place but is sufficient to scope the structure in as a potential resting place.

Searches for resting places often require the use of waders and may even require a boat in order to access some areas.

3.7.1 Special habitats

Reedbeds are favoured by otters, with larger reedbeds being more likely to be selected for resting (Kruuk et al. 1998). Radio-tracked otters in that study spent more time in reedbeds that were over a third of a hectare, although habitat selection is likely to depend on the broader context of which habitats are available. Reedbeds can also be used as natal habitat, with the otter using the reeds to form a natal holt structure (Taylor and Kruuk 1990). Any survey of a reedbed has the potential to disturb an otter in its place of rest and would need to be undertaken with care and to be justified. A survey licence may also be necessary. The presence of nesting birds should also be considered before venturing into a reedbed during the bird nesting season.

The same principles apply to finding resting places in coastal habitats. Additionally, coastal otters will rest in the intertidal range and these resting places are difficult to find with no prior knowledge of the locality. If such sites need to be identified, vantage point surveys will be necessary, although the methodology would need to be bespoke for each survey and carefully thought through.

3.8 Invasive surveys

Any sort of survey that entails illuminating the interior of a holt or viewing the interior using flash photography or endoscopic equipment is referred to as an 'invasive survey', because of the likely disturbance of any otters inside. It is

Figure 3.23 Habitat suggesting suitability as a resting place: (a) piles of flood debris, and (b) burrows under bridge supports.

crucial that due decorum is exercised when a surveyor is at a potential resting place. While a licence can legally allow for any resulting disturbance, it is always incumbent on the surveyor to avoid disturbance as far as possible, whether a licence is held or not.

3.8.1 Potential disturbance from invasive surveys

Firstly, in the UK at least, to cover you for any unforeseen circumstances, you should ensure that you hold a licence to undertake an invasive survey if it is required in that country. It is one thing to use a torch to rule out structures as part of a walk-over survey, but quite another if you are an (experienced) surveyor and think there is any chance that a structure could be occupied by an otter. In this case, you should immediately and quietly vacate the area. Any invasive inspections should be delayed until you are confident that the structure is empty. If there is an 'active' latrine (i.e. one that isn't completely weathered), an invasive inspection should not be undertaken but should be recorded as a resting place based on this field evidence. To determine whether a structure is occupied prior to your invasive inspection, a brief, two-to-three-day deployment of a camera-trap is necessary immediately prior to the inspection to determine that the structure is not occupied, so that you avoid being faced with a hissing dog otter or, worse, disturb a female with young cubs.

An invasive survey will deposit human scent all over the structure. You should limit this within reason if you are able, such as by wearing sterile gloves, and should consider the inspection as the only opportunity to look inside. Thus, it is essential that you are well equipped and know what you are looking for and how to find it. This avoids the need for repeat visits.

3.8.2 How to undertake an invasive survey

It is vital that an invasive survey is as thorough as possible. A strong torch may be all that is required at some structures to get a really good look. A phone torch or a small pen torch will be inadequate because the torch needs to be bright and have a good throw and spread of light. The ecologist may need to lie down and look up into the entrance to see inside. If a torch is inadequate, a series of photographs using flash may be sufficient as long as the whole of the interior is photographed. Discerning internal latrines can be tricky, especially if they are old and the floor of the structure is covered in loose material such as disintegrated bark. The next approach is to use a beefed-up endoscope. A typical bat endoscope will be inadequate because the light throw and field of view will be too small. We have used a home-made assemblage using a CCTV-type camera with LEDs on a pole wired to a car reversing screen and a battery pack. Action cameras placed on the end of a pole, with a wireless link to a phone or tablet, can be used in a similar way. These are tricky to use, but we have managed to view 6 m into a holt this way.

3.8.3 What to look for in an invasive survey

You are looking to confirm resting place status by finding any field signs associated with resting, such as an otter nest or a latrine, or to classify the site as a potential resting place by assessing whether the structure is large enough to house one or more otters, in which case you have a potential resting place. Some resting places can be flooded on occasion, and so while you ideally want to see a dry interior, this may not be the case if the river has been in spate or adjacent ground has been flooded recently if the structure is in or near a riverbank.

Structures with tunnels that have turns are usually unsuitable for endoscoping, but having a proper look may show the tunnel to be blind or too small for an otter and so can be ruled out. If you cannot see that it is blind or too narrow, a well-used tunnel may show smoothed sides and a smooth, padded-down base, meaning it could be a potential resting place.

3.9 Decision tree for identifying resting places

Figure 3.24 introduces a decision-making flow chart to guide a surveyor on how to go about assessing a potential resting place. It draws on our research and outlines

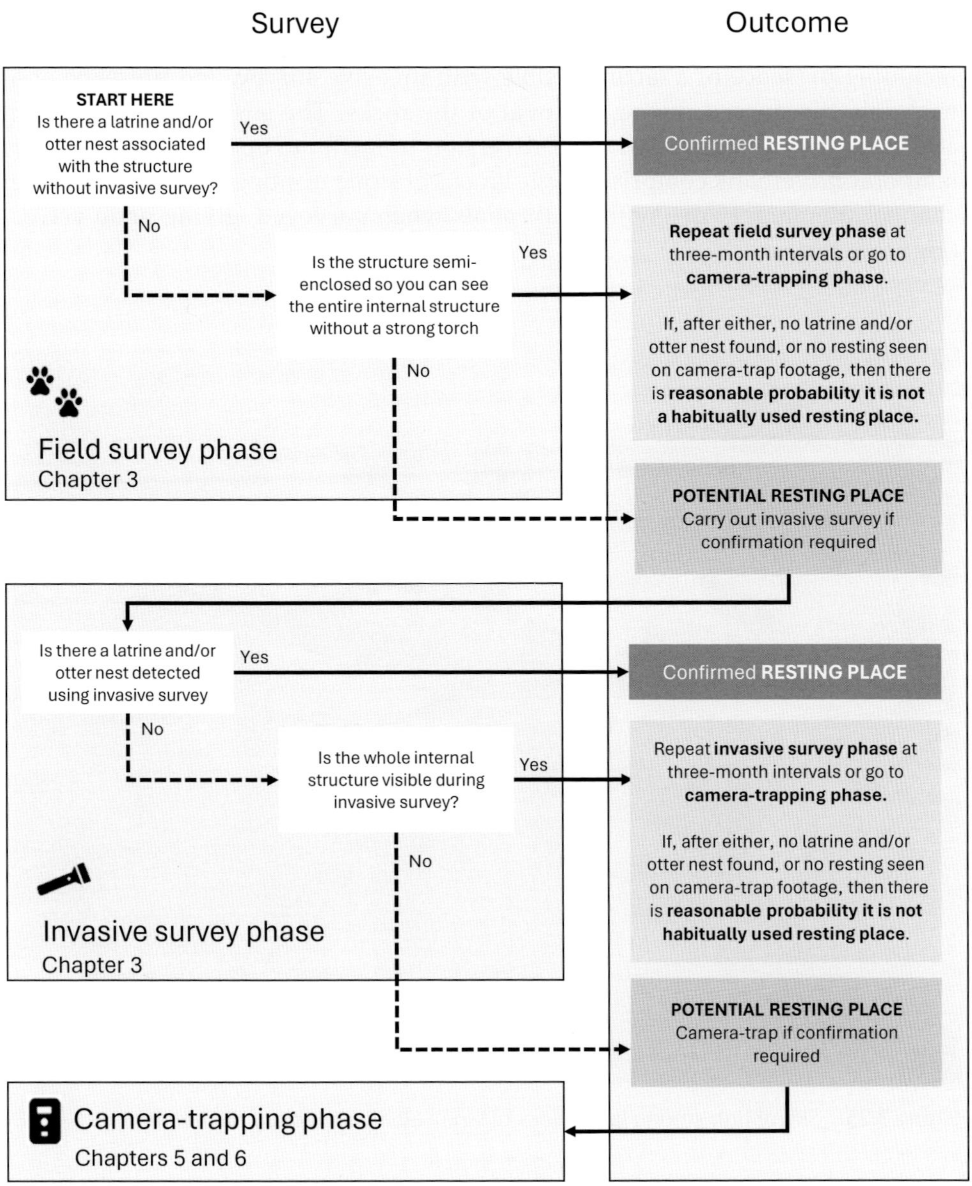

Figure 3.24 Decision framework for survey and categorisation of potential resting places.

how you need to look at and assess a structure first using field signs, and when camera-trapping should be deployed. A certain amount of professional judgement is required and this flow chart is not meant to be a rigid approach. Some potential resting places may be confirmed or discounted as resting places relatively easily. Others, however, may sit in the grey area such as an open site without a nest or a latrine; the surveyor should not discount these completely and in such cases repeat visits to check for a latrine and/or a nest, or camera-trap survey, should be considered.

3.10 Case study of a holt destroyed under licence in 2010

This case study further illustrates some of the aspects covered in this chapter, including presence of a latrine visible with invasive survey, and presence of a nest only discovered during removal of the rocks. The site was a temporary, artificial channel created for river diversion to facilitate opencast mining. The temporary riverbank comprised large blocks of stone fronting an earth bank. The channel was largely dry with small pools when surveyed, giving good access (Figure 3.25).

Figure 3.25 Site of holt among large block of stone along temporary riverbank.

Only one or two inconspicuous spraints were present on the ground in front of the rocks to indicate any presence of otter, which were found during a walk-over survey without any invasive inspection. Using a torch, a large latrine could be seen tucked away between two blocks of rock (Figure 3.26). The latrine contained the remains of uncountable droppings washed through and disintegrated, and fresher ones on top. At this time, latrines were not recognised as a field sign in their own right and this was interpreted as a 'spraint site', although it was recognised to be significantly large.

Removal of the rocks revealed the structure of the resting area (exposed in Figure 3.27). There was an earth base, boulders to either side and large slabs of rock forming the roof, which was very low, possibly as low as 20 cm. It was not clear where the entrance was. The bedding comprised rushes, grasses, sedges and moss, reflecting the local vegetation around the holt.

The nest itself when completely uncovered was approximately 70 cm diameter and completely dry (Figure 3.28).

Figure 3.26 Large latrine visible between the rocks, which had weathered to leave a spread of undigested remains such as fish scales and bones.

Figure 3.27 Resting area visible after partial removal of surrounding boulders.

Figure 3.28 Large (70 cm) nest visible constructed with nearby vegetation. This was found when surrounding boulders were removed.

3.11 Key points

- We use the term 'breeding place' to encompass natal resting places and maternal resting places. In other words, any of the resting places used by the female to give birth and raise the cubs until they are six months old. A 'breeding site' includes all of the breeding places and the supporting habitat.
- It is not usually possible to distinguish a breeding resting place (i.e. natal, maternal) from other resting places based on field signs.
- Presence/absence of spraints at a potential resting place is not a reliable indication of whether a structure is a resting place or not. Likewise, our research did not show a relationship between the number of spraints at a potential resting place and whether it was used for resting or not.
- The presence of a latrine and/or an otter nest is a good indication of a resting place. However, these signs are relatively rare and may be inside a holt and not visible.
- Resting places need an entrance of 10 cm or more. The sleeping areas are generally a minimum of 50 cm diameter to fit a resting otter. Smoothing of the substrate of any associated tunnel can also be noticed.
- Invasive surveys may require the surveyor to be covered by a licence to disturb otters. A potential holt should always be camera-trapped prior to an invasive survey to ensure it is not occupied at the time.
- Camera-trapping is often the only way to confirm that a potential resting place is used for resting and hence a protected resting place.

4. How camera-traps work

4.1 Introduction

In some cases, camera-traps may be crucial for monitoring potential resting places to confirm whether they are resting places and/or identify whether they are being used by breeding females. To effectively use camera-traps and be able to adapt your camera-trapping approach to the wide array of sites and contexts you might find potential resting places in, it is very helpful to establish an understanding of how they work and how this affects detection of otters. This will give a solid basis on which to carry out robust surveys, providing a foundation for the more practical guide to camera-trapping otter resting places in the next chapter.

4.2 Passive infrared detection

Camera-traps typically only record or take images when triggered (hence the name 'trap'). In theory they only record short videos or take pictures when an animal is in front of them. However, this system is by no means perfect, since a camera-trap may sometimes fail to trigger in response to an otter passing it or may be triggered by things that are not animals. Passive infrared (PIR) detection, which is used in many familiar movement sensors such as those for burglar alarms or motion-activated light switches, remains the primary means by which off-the-shelf camera-traps are triggered. While other bespoke trigger mechanisms have been tested on otters, such as physical switches triggered by body weight (Lerone et al. 2015) or actively powered infrared beam counters (Garcia de Leaniz et al. 2006), PIR triggering allows the camera-trap to be a single, self-contained, compact unit and to be highly energy efficient.

A basic understanding of the PIR trigger mechanism and how it relates to what is and isn't captured in camera-trap footage is key for getting the most out of your camera-traps. A very helpful description of how PIR detectors work has been provided by Welbourne et al. (2016). To summarise, all objects including otters and their physical environment emit infrared (IR) radiation (i.e. radiated heat). This is focused through a Fresnel lens (see Figure 4.1) onto a pyroelectric sensor. The sensor detects changes in infrared energy, for instance when a heat-emitting animal walks in front of the camera-trap and sends a signal to the camera to take a photo or video footage. Equally, if blocks of ice could walk, and one passed through this same area, this would elicit a signal to the camera. While it is often an animal which triggers the

Figure 4.1 Some key external features of camera-traps shown on three models from two brands. Numbered features are: (1) Fresnel lens in front of passive infrared (PIR) sensor (note the third model has two additional side-facing lens/detectors); (2) infrared LEDs, which in some models are visible and in others are behind a filter to ensure any red glow is not visible to the animal.

PIR sensor, movement of vegetation (e.g. in the wind) or rain can also trigger the PIR sensor because the vegetation or raindrops may have a different temperature to surfaces behind. The area over which the PIR sensor monitors changes is called the detection zone.

The primary benefit of PIR detection, which outweighs its limitations, is that it has a low power demand, meaning that between trigger events camera-traps are not expending much battery power. This differs from some continuous recording technologies that can detect movement but are comparatively power hungry. For example, many CCTV systems now have movement sensors, but they operate by noting when the pixels of the recorded image change by a threshold amount. To operate at night, they also need to continuously illuminate the target area, which uses additional power. Small battery-powered action cameras can be used to record wildlife but, because they record continuously, they only last for hours, in contrast to a camera-trap that can potentially be left functioning in the field for weeks.

4.3 Dimensions of camera-trap fields of view and detection zones

The field of view of most camera-trap models is generally similar in terms of horizontal angle. Looking at a representative range of cameras sold on trailcampro.com at the time of writing (we arbitrarily took the first camera from 14 brands) the mean horizontal angle was 47° and all except one fell within the range 38° and 54°. The exception was a camera with a wide-angled lens of 80°. We have found that in the footage of wide-angled camera-trap models it can be harder to recognise animals at greater distances (Findlay et al. 2020). This is illustrated in Figure 4.2 in which

Figure 4.2 Comparison of field of view of a wide-angle and normal lens camera-trap placed in the same location.

an otter toy 3 m away appears much smaller when using a wide-angle camera-trap compared to a standard camera-trap in the same position, and the proportion of the screen taken up by the toy at that distance is reduced. It should also be noted that the view of the camera is not just determined by the camera lens, but also by what obstacles are in front of the camera-trap that may obscure an animal, such as vegetation or undulating ground.

The dimensions of the detection zone of a camera-trap are determined by the Fresnel lens. Many Fresnel lenses on camera-traps are actually multi-zoned, meaning the lens is more like an insect eye made up of multiple smaller lenses for better performance. It is also useful to think about the detection zone in three dimensions (width, depth and height) rather than just two, in particular the ability of the camera-trap to detect animals at a lower angle than the camera. In the case of otters, which are quite low to the ground, it could be possible for them to pass very close the camera-trap without triggering the PIR sensor because they are too low – this is often called the 'sneak zone' (see Van Berkel 2014). Detection zones vary widely between camera-trap models in terms of number (i.e. how many zones the Fresnel lens has) and where they sit in relation to the field of view.

4.4 Camera-trap illumination

So that camera-traps can record footage in low light or at night, infrared LEDs are typically used to illuminate the field of view. These can be seen on the camera-traps in Figure 4.1. The IR used in the illumination is on a different part of the electromagnetic spectrum than the IR the PIR sensor detects; while the PIR detects at about 8 to 14 µm, the LEDs emit IR with much shorter wavelengths much closer to visible red light, at either 850 nm or 940 nm. In fact, because they are close to visible red light, 850 nm IR LEDs glow red as they omit both IR and visible red light and so are often called 'low-glow' LEDs. The benefit of these is that they provide stronger illumination for getting images or footage at night. However, from our experience these low-glow LEDs have the potential to disturb otters because they are likely to be aware that the camera-trap has switched on. The longer wavelength 940 nm LEDs are often called 'no glow' because they emit a slightly longer wavelength and so are not visible (at least to humans). We have used no-glow (940 nm) camera-traps extensively and otters do not appear to react to their illumination, compared to some reactions we observed to low-glow cameras we tried in the early 2010s, when camera-traps were a relatively new technology. This is especially relevant when trying to avoid disturbance when camera-traps are pointed at otter resting places.

4.5 A story to explain camera-trap detection

Now we understand a bit more about the PIR sensor, we should consider that a camera-trap 'sees' the world in two ways: (i) through focusing of far infrared radiation onto the pyroelectric elements (the PIR sensor), and (ii) through visible light and/or near IR focused via an optical lens onto an image pick-up device (the digital camera). It is the interaction between what causes the PIR sensor to trigger, how quickly the digital camera can react to that trigger, and the extent of the field of view of the digital camera itself that are critical to determining camera-trap detection, such as that of an otter. This is a complex process and so we'll try to illustrate it with a story.

Imagine two old friends sitting in a bird hide facing a wetland and waiting all day for a rare bird to appear. The first friend (the PIR sensor) is very dedicated. She is constantly alert and on the lookout. However, she has forgotten her glasses and is only vaguely aware of what is moving in front of the hide – it's all a bit of a blur. The second friend (the digital camera) is not so vigilant and snoozes the whole time. This second friend has fantastic camera equipment but relies on the first friend to nudge him awake when something comes along. His photo camera is quick and he just has to press a button to take a snap. His video camera can film some nice behavioural footage but takes a short while to turn on when he wakes up.

Imagine that a bird arrives. For the first friend, with her poor eyesight, we suspect that if the bird is close, big or straight ahead, or if it is brighter compared to the dull background (in the case of PIR that typically means warmer relative to the background), there's a good chance she'll spot it. But if the bird is relatively still, far away, smaller, off to one side in her peripheral vision or is a similar colour to the background (for PIR sensors this means a similar temperature), there's every chance she won't see it. If she thinks she can see something, she nudges her friend, the digital camera, awake so he can get a record of it.

The friend will take a moment to stir awake, and then a moment to grasp a camera; depending on whether they had pre-agreed to take photos or videos of the rare bird, he will be ready slightly faster (the photo camera) or slightly slower (the video camera). Without looking, he raises his camera, points it out of the hide and starts recording. With luck, he catches an image or footage of whatever it was that alerted the first friend. For him to take footage of the rare bird depends on several things: (i) the bird is there, (ii) the bird is close enough, large enough, distinct enough or moving enough for the first friend to have sensed it (despite her limited vision), and (iii) it has not flown out of view by the time he wakes up and starts snapping or filming. There are also other possibilities that might lead to the pair not getting the footage of the rare bird, including (iv) on later inspection of the footage it turns out to not be the rare bird after all, just a different species, or (v) the movement that caused the first friend to react wasn't a bird at all, but wind blowing a carrier bag across the wetland.

This story is intended to illustrate some important concepts in camera-trap detection. The first friend's perception is analogous to the detection of a PIR sensor, in that it can easily miss things (particularly if they're small, far away, still or too similar in temperature to the environment), yet when it does get triggered it does not discriminate between movement of something like an otter, and movement of vegetation in the wind. The view through the second friend's camera is analogous to the field of view of the camera, in that it has potential to take great footage but is totally reliant on the PIR to tell it when to start, and its reaction is delayed. For a camera-trap to detect an animal of a particular species such as otter (and by that we mean to generate recognisable footage), all three scenarios (i), (ii) and (ii) above need to have occurred (i.e. the otter has to be there, it has to trigger the PIR sensor and it has to still be in the field of view before the camera turns on) – the failure of any will mean that otter footage is not captured (Findlay et al. 2020). In addition, scenarios (iv) and (v) are analogous to gaining footage of non-target species and false triggers respectively, both of which mean more battery drain and more work checking the footage.

4.6 Detection probability in otters

Now that we understand a bit more about camera-trap detection, detection zones and fields of view, we will discuss detection probability specific to Eurasian otters, based on several pieces of research. Detection probability is the probability that if an animal such as an otter is present in the environment (and thus available to be observed/recorded) we record footage of it on our camera-traps. The overall process of detection in camera-trapping consists of a series of smaller, sequential processes that must happen for the camera to record useable footage. Let's say that one of our survey goals is to know if a holt is occupied by a female otter that is pregnant or lactating. To determine this the otter must be present in the area and to trigger the camera-trap via the PIR system. This may sound like a given, but a variety of factors can reduce the chance of the PIR system triggering, even if the animal is present, such as closest distance of the otter to the camera-trap and the temperature (of the otter's surface or the background environment).

Once an otter has triggered the PIR, the camera footage must register the otter itself in the footage. Again, this may sound like a given, but there is always a delay between a camera-trap triggering and the initiation of recording (we call this the 'trigger time') and also a delay between the completion of one image or video and the next trigger. The trigger delay is included in the programming and the minimum is usually 1 second, although some camera-traps will continue to record video as long as the

PIR is sensing movement; this ability may be limited to daytime footage. As otters can be quite fast (like many mustelids), they can easily have run out of the field of view by the time the footage starts. We know this happens because: (i) sometimes all we see is the end of an otter tail flick out of shot, and (ii) in our research with captive otters we found by using CCTV that an otter had passed and triggered a camera-trap, but the footage missed the otter (Findlay et al. 2020).

Finally, if we do get footage of an otter, the footage has to be of high enough quality to see the animal sufficiently clearly for our survey goals. To identify a pregnant or lactating female may mean capturing footage from a good angle to see the belly or teats, while for other study goals it may be sufficient just to see that the animal is an otter (of any sex or age). Thus, 'quality' here doesn't just mean the number of pixels the camera has, although that helps, but also where the otter is placed in the frame, the angle of view (in some cases a lower angle might be better for identifying sex), or how close (it may be out of focus) or distant (less detail is visible) it is.

4.6.1 What drives trigger probability in otters

In our study of captive Eurasian otters in Devon, where we knew from our CCTV that otters had passed in front of camera-traps, we found that trigger probability (the probability of the PIR triggering the camera-trap, regardless of what the footage ends up showing) fell sharply after a few meters and was about 50% at about 5–6 m for dry otters. In a study at a wild otter holt where we did not separate trigger or registration probabilities (but trigger probability will still be a substantial component), a camera placed at 1.6 m from the entrance detected otters about 30% more often than one placed at 4.2 m from the entrance (Findlay et al. 2017). This is because the heat (IR energy) emitted by an animal dissipates according to the inverse square law (Figure 4.3). This means that the energy received by the sensor at 2 m would be

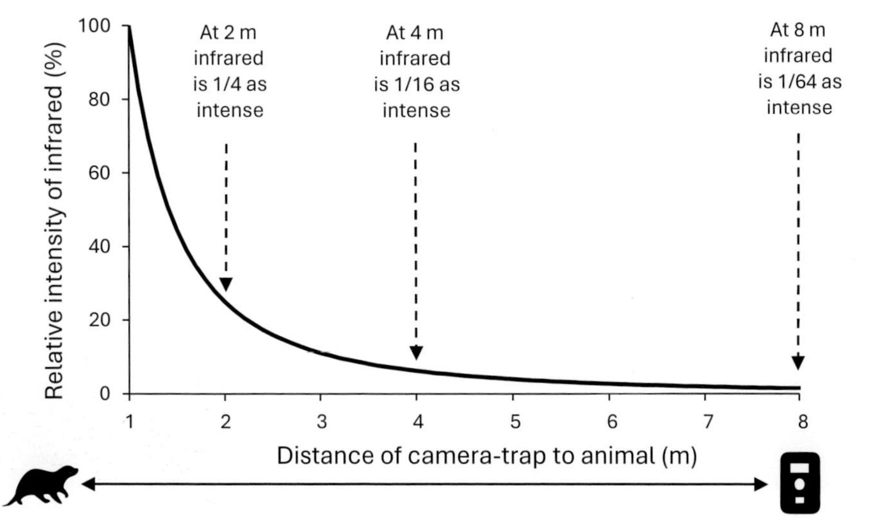

Figure 4.3 How the intensity of infrared radiation declines with distance from an animal's surface according to the inverse square law. Infrared radiation emitted from an animal can trigger a camera-trap, but the intensity of that infrared declines proportionally to the square of the distance.

one-quarter of the energy received by the sensor from the same animal at 1 m, and at 3 m the sensor would receive one-ninth of the energy it would receive from the animal at 1 m. Therefore, at 10 m, the sensor receives one hundredth of the energy that the sensor would receive at 1 m. All of this assumes a complete lack of intervening vegetation or other blockages to the energy.

Wetness of coat is also a big factor affecting trigger probability of otters. As aquatic hunters, and with resting places often close to water, they are frequently seen on camera-trap footage soaking wet (Figure 4.4). In terms of what we know from PIR

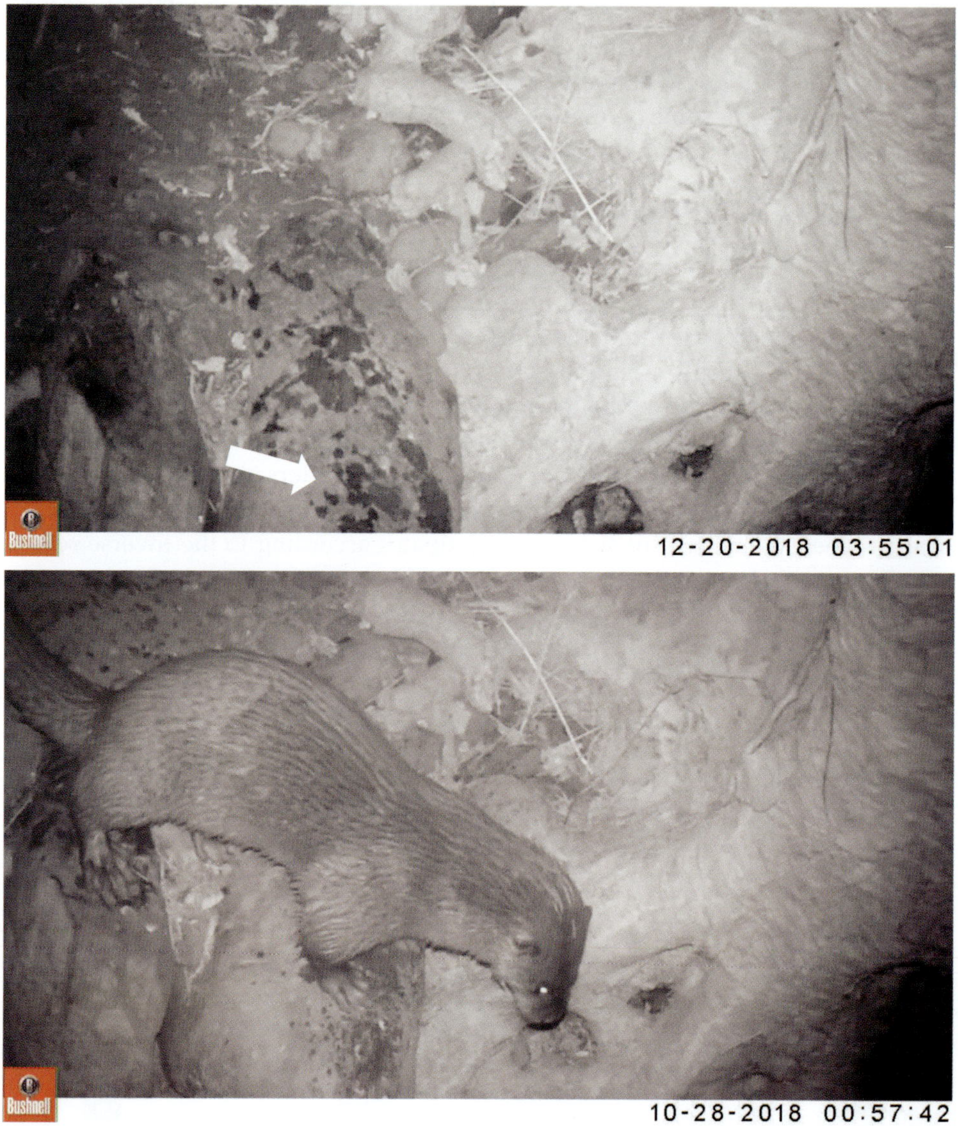

Figure 4.4 Example of failed registration in camera-traps. The top image is the first frame of a camera-trap video where an otter was missed. The wet footprints (some clear ones indicated by the arrow) indicated that the otter had just passed but was too quick for the camera-trap. The lower image shows a wet otter leaving footprints at another time when it was captured on camera.

detectors, which are triggered by a contrast in temperature between animal and background, a wet coat (which is often colder than when dry) can impact trigger probability. This phenomenon with otters was identified by Lerone et al. (2015), who noted that a spraint site on a boulder protruding from a river was being regularly sprainted on by otter, yet the otter was not being captured by a PIR camera-trap trained on the site. This effect of reduced trigger probability does diminish as the coat dries, but this could take some time, which will depend on the conditions. For example, we found in a trial that it could take 30 seconds or so for the coat of an otter emerging from water to reach an approximate 3°C difference in temperature from the background environment (Findlay et al. 2020).

In the same study we found that trigger probability of wet otters was substantially lower than that of dry otters and fell more quickly as distance increased. This means that when as close to a camera-trap as an otter could be, trigger probability was only predicted to be 80% for wet otters (as opposed to 100% for dry otters). In addition, trigger probability fell to 50% at about 3–4 m for wet otters, as opposed to 5–6 m for dry otters. Similar phenomena have been seen in other semi-aquatic mammals such as Eurasian beaver *Castor fiber* (Coxon 2017; Martin-Walker 2017) and Asiatic small-clawed otter (Duggan 2016; Roche 2016).

4.6.2 What drives registration probability in otters

Registration probability in some ways is less intuitive than trigger probability, since it is conditional, i.e. it is the probability that an animal will be recorded in the camera-trap footage given that the camera-trap has already triggered. Registration probability can increase the further the animal is from the camera-trap, particularly if the animal is moving quickly across the field of view (Findlay et al. 2020). This phenomenon is explained diagrammatically in Figure 4.5. If a quick-moving animal is close to the camera-trap when it triggers, it takes less time for it to cross and leave the narrower part of the field of view close to the camera-trap. Conversely, if it triggers the camera-trap but is further away, it will take more time to cross the wider part of the field of view and so is more likely to still be visible when the camera-trap starts recording.

It is worth understanding the concept of trigger probability and registration probability, since it may help to explain lots of blank footage when otters could be running straight past your camera-trap. One thing to bear in mind with registration probability is that for otters it was highest when the animal was walking compared to running or trotting, and particularly high when the otter 'loitered' for a while in the field of view (Findlay et al. 2020). So, at any spot where an otter might pause or slow down momentarily, for example a spraint site or where they have to change direction sharply, the registration probability may be higher than at a site where the otter can run in a straight line across the field of view.

4.6.3 What drives footage quality in otters

Capture quality (the usefulness of the videos/images obtained of otters) very much depends on the aims of your monitoring. In some studies, the ability just to identify otters may be sufficient. In the context of monitoring potential resting or breeding

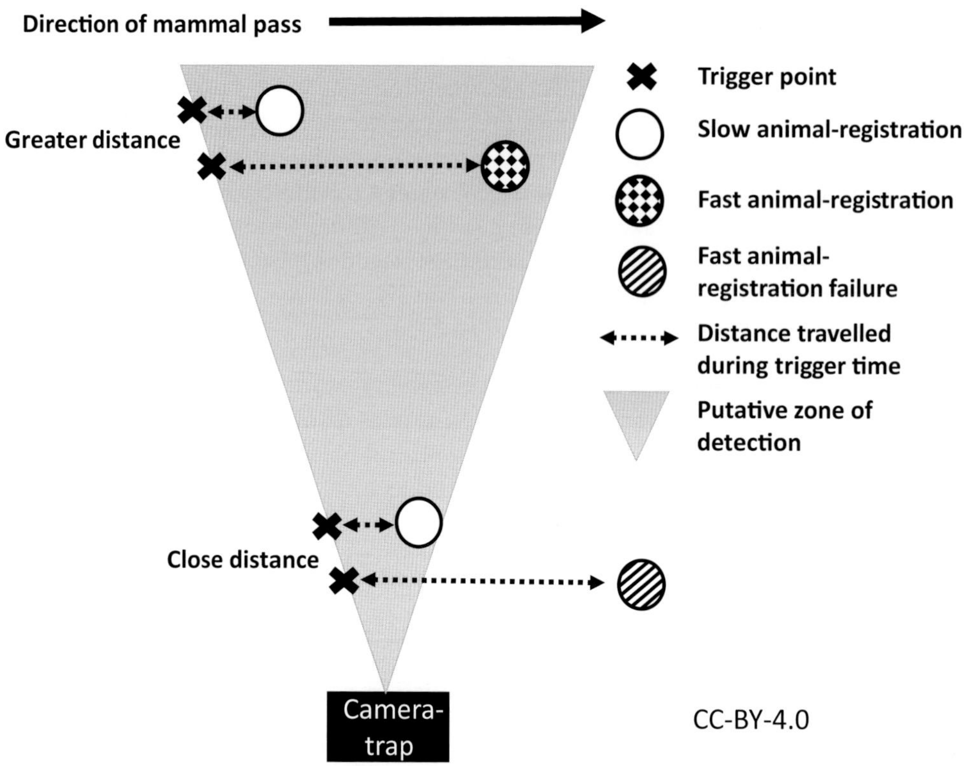

Figure 4.5 Hypothesised mechanism showing how distance to camera-trap can interact with animal speed to influence registration probability. Registration probability is positively affected by distance due to the larger area within the field of view at greater distances. Conversely, faster moving animals can completely pass through the small width of the field of view close to the camera-trap before the camera starts recording. From Findlay et al. (2020) (CC BY 4.0).

places, however, other information such as sex, age, breeding status, number of otters and behavioural footage may be crucial, so capture quality does matter.

In our long-term study of a holt in Scotland, where we placed one camera-trap at 1.6 m from the entrance and another 4.2 m away, the closer camera-trap was about 30% better at detecting presence of otters than the distant camera, but was about 53% better at producing footage that would allow identification of sex (Findlay et al. 2017), because it gave more close-up video clips from which primary and secondary sexual characteristic could be assessed. As discussed above, camera-trap height is important for several reasons, including detection probability, but the height and angle of the camera might also be important for quality of footage in the context of otters and their resting/breeding behaviour. For example, to detect whether a female otter is pregnant or lactating, or to detect the presence/absence of testes of an adult male, a lower viewing angle is important.

4.7 Key points

- Understanding how camera-traps function is important for getting the most out of them during your otter surveying.
- Camera-traps are triggered via passive infrared (PIR) sensors, which have the benefit of being highly energy efficient, but the limitation of relying on difference in surface temperature between an animal and the background, and also the animal remaining in the field of view for long enough for the camera to start recording.
- The PIR sensor's detection zone, the camera's field of view, and how the PIR sensor and camera interact are key to what camera-traps detect and what they miss.
- Camera-traps can record footage of otters at night by illumination with infrared LEDs, but the type of LED can influence whether the otter can see that illumination or not. No-glow LEDs are essential to avoid disturbance to otters at resting places.
- Camera-traps are more likely to trigger when otters are closer and when they are dry. If an otter does trigger the camera-trap, footage is more likely to be recorded if the otter is loitering or moving more slowly so it doesn't leave the field of view before the camera turns on.
- For otters, sex and age characteristics are easier to see when the otter is closer and the camera-trap is placed at a lower angle.

5. Deployment of camera-traps for otter surveys

5.1 Introduction

As discussed in Chapter 3, an otter resting place can be confirmed with a reasonable degree of confidence if there is presence of an otter latrine and/or an otter nest. However, in complex or deep, enclosed structures, these features are not always visible on inspection, even via an invasive survey. Where an experienced surveyor deems that the structure seems suitable for otters, we would consider it to be a potential resting place. This status may be enough to protect the site with a buffer zone during construction projects. However, where impacts are predicted from construction or other activities, confirmation or disproval as a resting place would be necessary to assess impacts and apply for a licence for those impacts. Furthermore, enhanced status as a natal or maternal resting place may need to be assessed. In such cases, camera-trapping may be crucial. In this chapter we will discuss key questions relating to practical aspects of camera-trapping at potential resting places and other camera-trapping scenarios. We look at choice of camera-trap, which settings to use and deployment in the field. We first introduce general considerations for camera-trap set-up and settings and then give more specific advice for camera-trapping (i) potential resting and breeding places and (ii) spraint sites and otter paths.

5.2 General considerations

5.2.1 Minimising the risk of disturbing otters

Camera-trapping has the potential to disturb resting otters if undertaken recklessly. Great care is needed to minimise this risk, but even with good protocols in place this risk can't be completely eliminated. Hence, a camera-trapping licence is likely to be necessary, in the four UK nations at least. Radio-tracking and camera-trapping studies indicate that otters regularly move from resting place to resting place and so it is possible that they do not have a fixed expectation of a resting places' surrounds. The river environment is dynamic, depositing debris and sweeping it away again and so the appearance of a novel object such as a camera-trap is not out of the ordinary; species such as badgers that occupy a sett almost continuously in a more static

Box 5.1 Code of conduct we follow during set-up, maintenance visits or take-down visits to potential resting places

- Keep mobile phones on silent.
- Use emplacements that enable camera-traps to be removed and replaced very quickly. Swapping a fresh camera-trap in and removing the spent camera-trap has been timed to take 20–30 seconds at an average site.
- Avoid knocking in stakes. Instead, use an auger to create a pilot hole. The stake can be quietly firmed in or wedged with small stones.
- If possible, place the camera-trap in front of debris or vegetation to disguise its outline.
- Avoid frequent visits to check camera-traps (three weeks is considered ideal; see section 5.6.3).
- Avoid touching the ground or resting place entrance (i.e. depositing human scent), especially any spraint piles.

environment may notice new features more readily. We consider that the presence and activity of the ecologist setting up by the structure and swapping camera-traps has more potential for disturbance than the camera-trap itself. Therefore, we try to minimise set-up time and follow a strict code of conduct (Box 5.1) in addition to reducing the frequency of visits to the camera-trap to change the batteries and SD card.

Following this approach we have found that all resting places identified in the Tweed catchment (Findlay et al. 2023) were used multiple times in the presence of camera-traps, which we visited every three weeks or so.

5.2.2 What camera-trap should I use?

Whether you already have a set of camera-traps or are looking to buy some, it's important to consider which features are needed to camera-trap potential otter resting places. Recommending technology in a book requires caution, since new models with new settings and features may be available at the time of reading that weren't available at the time of writing. The features discussed in Chapter 4, PIR detection, IR illumination and the ability to take video and/or still images, are features of virtually all camera-trap models. Despite this, there is a large range of camera-trap models available, varying widely in specification, flexibility of settings and price. Not all settings we discuss here will be available on all models, and in many cases it is best to decide which settings are important to you and buy a model that supports these, then learn what your camera-trap model can do and how it varies between settings through testing and trialling. Table 5.1 outlines some recommended features of camera-traps that would be suitable for otter surveys.

Table 5.1 Recommended features of a camera-trap for monitoring otters, particularly at potential resting places. Not all models will have these features, so it is worth comparing the specifications between models.

Feature	**Recommendation**
Trigger time	0.5 seconds or faster for video
IR illumination wavelength	Black IR at 940 nm wavelength is *essential* for camera-trapping of resting places. This is also called 'no glow'.
Ability to set duty time	Extremely useful to be able to 'sleep' the camera-trap during the main part of the day, but not absolutely essential.
Detection zone	Standard is about 40–50°, which is fine for most sites.
Standard or wide-angle lens	Standard field of view is about 40–50°, which is fine for most sites. Wide angle may be useful for very close quarters, such as at 1 m from a ledge or burrow entrance. At greater distances, wide-angle lenses are a disadvantage due to loss of details of the animal.
IR illumination adjustable	Ideally the model should allow the brightness of the IR illumination to be reduced. This is necessary to avoid over-exposure at close quarters and also saves power.
Battery settings/ compatibility	Rechargeable batteries are more economical and environmentally friendly but do not perform well in cold temperatures. In the UK, the chosen model should ideally allow the user to use programming settings to switch between the two. Rechargeable batteries can then be used for most of the year except for periods when temperatures are predicted to stay below zero, then non-rechargeables can be swapped in. It should be noted that rechargeable batteries may lose charge when not being used. Some brands, such as Eneloop, lose less energy when not being used and so last much longer in the field.
Wireless/cellular data transmission	At the time of writing, models that can transmit data mostly only do so with still images and can be unreliable where the signal is poor, so this function is not required.

5.2.3 Consistency of model

Studies that have compared multiple camera-trap models facing in the same direction with the same animals passing in front all show that camera-trap models differ in the extent to which they detect the same animals (Findlay et al. 2020; Randler and Kalb 2018; Wellington et al. 2014). While camera-traps will have different detection probabilities due to differences in hardware design and firmware features, if set up using the approaches we suggest, and if certain things are avoided, differences between models should be minimised and there is no reason why more than one model cannot be used. Probably more important than having the same camera-traps for every otter resting place you survey is using models you find reliable, have the features you require, you know how to work well, and do not have features that will disturb otters such as red-glowing LEDs or noise emissions.

5.2.4 Sensitivity settings

Some camera-traps will allow you to adjust the sensitivity of the PIR sensor. It's tempting to think that maximum sensitivity is always better as it will increase the chance of triggering by otters, but it is important to note that with maximum sensitivity the camera-trap will, in turn, be far more sensitive to other things such as waving vegetation, small mammals at night or birds during the day. In some circumstances, if you are getting many false triggers, and there is not a safe way of reducing vegetation movement (such as trimming) or other factors, reducing sensitivity may be an option. This aspect has not been studied in detail with otters, but we have typically used the default or medium sensitivity setting of the camera-traps used.

5.2.5 Still images versus video footage

Most camera-trap models have options to take either still images, video clips or a combination of the two (for example, one or more images followed by a video clip, called 'hybrid mode'). Images can also be programmed to occur in rapid bursts. These settings will vary between models and it is worth looking at what is possible. There are different reasons for taking images rather than video. Images use up less memory storage and power and can be reviewed relatively rapidly. Where the purpose is just to confirm presence of otter, images may be preferable, although in most UK river systems and some coastal areas otter presence can either be assumed or else more easily confirmed by field sign searches. In the case of monitoring potential otter resting or breeding places, however, we recommend the use of video clips for most scenarios; they capture more useful information relative to resting and breeding, such as physical characteristics used to try to identify age or sex, or to record behaviours associated with resting and/or breeding (explained in Chapter 6).

If you have previously used images and then change to video clips you should bear in mind that this may increase the trigger time, especially with older camera-trap models. Trigger time is the time between an animal triggering the camera-trap and an image being taken or, for videos, the first frame being recorded. Trigger times are typically shorter for still images than for videos, and the value advertised for a camera-trap is invariably the lower of the two, although note that this disparity seems to be declining with newer models. A longer trigger time is expected to result in more missed registrations because the animal has more time to leave the field of view before recording begins. This needs to be considered if you get quite a few instances of blank footage, and it may be worth checking the trigger time on your camera-trap (method given in section 5.6.1) to check it is not overly long, ideally about 0.5 seconds or less for otter. Nevertheless, in the case of monitoring potential resting places of otters, the benefit of better information obtainable from video clips is clear, so video is recommended. The trigger time on many models has been tested by www.trailcampro.com, which publishes the results of rigorous tests on its blog site; you can check this before buying.

5.2.6 Length of video clips and delay between clips

Where video clips are selected, duration of video has to be specified as part of setting up. Models we have used allow videos ranging from 1 to 60 seconds. The choice is a balancing act. Too short and any extended behaviour will be recorded as brief snippets with time gaps between them (depending on any trigger delay settings you have selected); too long and every trigger, including every mouse that dashes past or waving vegetation, will create a large data file that you may have to watch in case it contains otter footage.

In an analysis of video clips taken at a known natal holt we monitored for several years, we initially recorded videos for 30 seconds, selected relatively arbitrarily. However, when we looked closely at these video clips we usually saw the otter characteristics needed to identify its sex (either using primary or secondary sexual characteristics) and found that in theory we could have reduced video clips by about 10 seconds and still recorded the vast majority (95%) of useful information (Findlay et al. 2017). A reduction from 30 to 20 seconds would generate a substantial saving in terms of memory storage, power consumption and analysis time, and it was felt that this outweighed the effect of losing 5% of information, and so we have used 20 seconds as our standard video duration in our research (Findlay et al. 2023) and consultancy work. It should be noted that this optimal clip duration may vary between sites and contexts, but it is worth considering adopting a similar approach or at least asking yourself critically how much new information you get from the final part of your own video clips. If the answer is not much, you could consider reducing their length.

Another setting to consider is the trigger delay, which is the time during which the camera-trap is programmed to be dormant between clips. Again, this can be a balancing act. For example, if you selected no delay between clips, a 30-minute windy period of waving vegetation would almost all be recorded and would consist of much otter-free footage. On the other hand, if you had three otters entering a resting place within a few seconds, and leaving the resting place five minutes later, using no delay between clips would be ideal because all otters would be likely to be fully captured. Conversely, if the delay were to be set to 30 minutes, waving vegetation might not be a problem anymore, but you might only capture one otter going into the resting place (if that), because the camera-trap would be in idle mode while the other two otters followed and again when the three otters emerged. Your data would be significantly affected. On balance, our recommendation for monitoring potential resting places or any other site where you want as complete an account of activity as possible, is to set the delay between clips to the minimum for the model, since every pass of otter may be crucial because it may reveal a resting event.

5.2.7 IR illumination type and level

For camera-trapping at resting or breeding places it is essential that the LED illumination is black IR (940 nm wavelength LEDs). This should be clear from the documentation for your camera-trap but can also easily be tested in a dark room or outside at night, because red IR (850 nm LEDs) is visible, particularly when viewed face on. Manufacturers will describe their camera-trap illumination as black IR, 'covert' or no glow but you should *always* check the wavelength of IR as these terms are not specific. Red IR (850 nm) cameras can cause reactions by otters (Figure 5.1) and, although this could be OK for camera-trapping along runs or at spraint sites, it is not appropriate near resting places and there is a large risk it could affect how otters use resting or breeding places. Another consideration is that red IR cameras are more visible to potential thieves, who may see them glow at night, and such theft may cause disturbance to a site and lost data.

When checking your camera-trap for any red glow you should also confirm that it is not emitting any audible noise. This is best done in a dark, quiet room. There are several reasons why cameras may unintentionally emit noise (Meek et al. 2014), which may be high-pitched sounds while the camera-trap is on or a click that may relate to an IR filter engaged during daylight or disengaged during low light.

Most camera-traps allow you to adjust the brightness and hence the distance that the IR LEDs can illuminate. The effect of this brightness level cannot be seen with the naked eye but becomes clear when looking at footage that is overexposed and suffers from a 'white-out' effect. This can occur if the backdrop such as a riverbank is close to the camera-trap, or there are items such as rocks, branches or vegetation in the foreground that reflect back the IR illumination at high intensity. When camera-trapping potential or known resting places, the lowest brightness setting is always recommended because, if you follow our recommendations below, your camera-traps will be set close to the target capture area. This also has the added benefit of reducing power consumption. In these cases it is essential to test the illumination prior to setting up in the field to make sure that there is enough illumination for a good image.

5.2.8 Setting duty time

Some camera-traps can be programmed to set a period of time when the camera-trap is dormant and a period when it is on duty, sometimes called the 'duty time'. As otters are predominantly nocturnal, this gives you the ability to sleep your camera through the daytime and avoid capturing birds, waving vegetation and so on. We found that 89% of otter activity at a holt in freshwater habitat at our long-term study site was between one hour before sunset and one hour after sunrise, i.e. nighttime plus a one hour buffer each covering dusk and dawn (Findlay et al. 2017). You should be mindful that sunrise and sunset change slightly each day, substantially if your camera-trap runs unattended over several weeks or months, and so you need to think ahead to the end of your intended camera-trapping period. If daylength is shortening, you need to make sure that the important sunrise and sunset periods are

Figure 5.1 Examples of reactions to camera-traps using red IR illumination from the early 2010s: (a) otter cub staring at the camera-trap, shortly after which it crawled low along the ground past it; (b) otter in posture to get a better look; it spent some time going back and forth staring at the camera-trap. These contrast with the lack of reaction to black IR camera-traps we have observed since.

fully captured. Interestingly, at our long-term study site, there was increased daytime movement in and out of the resting place during one natal period, contrasting with others. So, if you suspect that you may have natal activity, it is best not to set a duty time. In coastal situations the activity of otters is less governed by daylight and so activity could be expected both day and night, so setting a duty time is not recommended as valuable activity may be missed.

5.2.9 Daylight savings time

In the UK, midnight on the last Sundays in March and November causes much head scratching because the clocks gain or lose an hour for daylight savings. Otters, clearly, are blissfully unaware of this. You can choose to go with the flow here by changing the times of your camera-traps in time for these changes, or, as we do, ignore daylight savings and programme the date and time to a standard time zone such as Greenwich Mean Time. This may mean that for half of the year, your date/time stamp is an hour different from the metadata on your video file. It is important to be aware of this, especially if you are setting duty time.

5.3 How should I deploy my camera-traps at potential holts?

5.3.1 Multiple camera-traps at a potential holt

Camera-traps, as useful as they are for monitoring, can be unreliable in terms of malfunction, missed detections and depleted battery and/or memory if they get a lot of triggers (particularly due to vegetation or other nuisance triggers). In addition, they can only record (and detect) within a restricted horizontal angle (typically 40–50°), which is a challenge if a structure is in complex terrain or has multiple entrances. For optimal detection, we advise setting two camera-traps per entrance of each potential resting place. This is more important if you think you will have impaired detection probability, for example if you are camera-trapping otters shortly after they exit from the water. If there is a site where disturbance by an ecologist is anticipated to be an issue, but theft or flooding is not, placing a pair of cameras each anticipated with good detection but with one set to record during the first part of the night and the second to record the second part of the night, visit frequency to maintain the camera-traps (discussed below) could be reduced. With a pair of camera-traps set at different angles, sometimes one camera-trap will collect good footage and sometimes the other, depending on the activity and path of the otter(s). This way, the camera-trap angles can complement each other, bringing more information than if there were just one.

5.3.2 Distance, height and angle of camera-traps relative to resting place entrance

In an ideal situation where there is flat ground and no other restrictions, such as in Figure 5.2, the two cameras should be placed at about 1–3 m from a structure's entrance, and at different angles, allowing potential for different views of the same otter(s), which can be useful for sexing and aging. This close distance is important

Figure 5.2 Two examples of camera-trap set-ups at potential resting places where there is relatively flat ground in front of a single entrance, such that there is a wide choice of camera-trap placement locations.

because trigger probability falls sharply with distance from camera-trap, an issue exacerbated when otters are wet (Chapter 4). Obviously, neither camera-trap should be on any otter runs or impede entry into the structure.

Because the wavelength of the IR that camera-traps emit for illumination (0.85 or 0.94 μm) is different to that which the PIR detects (about 8–14 μm), camera-traps cannot trigger each other if one is in view of the other. However, if one of your camera-traps is directly facing the other, at night it will cause glare that obscures any footage if both camera-traps are on, so this should be avoided. If the camera-traps are at an angle to each other this should not occur (Figure 5.3).

Distance is one of the key drivers of detection probability for camera-traps, due to its effect on triggering success. Therefore, we recommend that camera-traps are not placed too far from holt entrances because this will risk missing otter passes, or behaviours that may be crucial in identifying the function of the site. Each site is different, and an ecologist needs to work with the terrain available; we would recommend placing cameras quite close, for example 1–3 m (averaging about 2 m) from the entrance, while not blocking any obvious paths/runs.

In an ideal situation with flat ground (Figure 5.2), a low height and an angle parallel to the ground is recommended. A height of about 20 cm off the ground, with a flat angle towards the entrance or area in which the otter is expected to most often be, is most likely to yield views that will help determine sex/age or breeding status. If the ground slopes slightly the camera-traps should be angled to match the slope, to maximise detection. Facing camera-traps into low sun should be avoided because

Figure 5.3 An otter commuting through a tunnel. We can see the illumination of a second camera-trap, but the otter should not be able to see this, nor can it trigger the camera-trap we are looking through because the IR radiation from the camera-trap's LED flash has a much shorter wavelength than that which will trigger the passive infrared detector of a camera-trap. As the camera-traps aren't facing each other, there is no loss of detail due to glare.

the sun may trigger the camera-trap and/or cause glare on any footage, obscuring the image.

5.3.3 Dealing with challenging terrain, and using emplacements

Of course, potential resting places will frequently fall foul of the ideal situation illustrated above, and may be in complex, undulating terrain with tree roots, rocks, rank vegetation. The entrance may also be directly above water. This is where the surveyor needs to plan carefully. The golden rule of camera-trap placement at otter holts is that you put the camera where you think it will get the best balance of detection probability and viewing angle; you should compromise only where this would impede the passage of otters on their normal route. In some cases there may be something to attach the camera-trap to at exactly the right distance, height and angle, but frequently there will not be. Attaching your camera-trap to the nearest convenient tree or fencepost, or across the river because the entrance faces that way, may be the easiest option, but will undoubtedly compromise the quality of the data. The nearest convenient tree for the camera-trap could be metres away or at a suboptimal angle, and so have poor detection probability, which will invalidate the survey.

Assuming you will be monitoring for at least several weeks (see section 5.5.2) careful planning is essential. This is where emplacements come in. An emplacement is any additional item you place in the environment such as a stake, beam or bracket that creates the optimal location and, if designed well, allows you to swap out the camera-trap quickly during maintenance visits (to change traps, batteries or SD cards), with no modification in viewing angle. Some examples of emplacements have already been shown (e.g. Figure 5.2). Further examples are provided in Figure 5.4. One emplacement consists of a bespoke wooden box with sideways angle on a steep riverbank, which allowed the camera-trap to be easily swapped during maintenance visits while maintaining the same field of view. Another resting place had a dead branch in front of the entrance. The camera-trap was placed in a plastic pipe with a window cut out of it, to allow the camera to 'see' over the branch while also providing camouflage from would-be thieves.

In addition to bespoke emplacements, increasingly there are commercial alternatives to only using the strap that comes with most camera-trap models. Many camera-trap models have a quarter-inch screw fitting (of the type often used for spotting scopes or cameras) and there are various types of stand or fittings that attach to this, which may be useful in different contexts. However, note that some camera-trap stands such as single rods that stick in the ground can suffer from wind-shake and can be easily knocked over by animals (badgers!), so you should select emplacements that are sturdy.

5.3.4 The challenges of dense vegetation

Dense vegetation is a real challenge. If it lies between the camera-trap and the target capture area it reduces detection probability of the otter, obscures the image, may cause over-exposure and is likely to be the cause of hundreds of clips of endless

Figure 5.4 Examples of where bespoke emplacements may be needed when a potential resting place is on challenging terrain, which will not allow good camera-trap locations (in terms of distance, obstacles, angles or high risk of theft) using wooden or plastic pipe emplacements.

waving vegetation in breezy conditions. There are three options to mitigate this. The first is to do some subtle 'gardening', pulling up or pressing down patches of vegetation and/or placing a nearby log over it to keep it down. This is fine if only a light touch is needed and if you are certain the structure isn't a natal holt, where otters may be far more prone to disturbance. The second option is to raise the height of your camera-trap to look over the vegetation if the entrance to the structure is uncluttered. It's a reasonable and necessary compromise at some sites. Finally, dense vegetation can make some sites unfeasible for camera-traps in the summer, so the third option is to install a camera-trap when the vegetation is at a natural minimum. Forward planning is necessary in managing vegetation: a camera-trap set in April with the intention of leaving it for six weeks should be set at a height predicted to look over growing vegetation.

5.3.5 Protecting camera-traps from theft/damage

Damage and theft are an expensive loss and mean that any data collected to that point is gone. For otters, it's important to keep an eye on weather forecasts and predict potential river spates if your cameras are close to a watercourse, as they often will be. Camera-traps can survive immersion occasionally but there's a high risk of water damage or being washed away. In some cases, it may be good to remove cameras pre-emptively and replace them when water recedes, although this should only be done when safe to do so (i.e. prior to water levels rising significantly).

Theft is a more insidious problem, and camera-traps might also attract people towards a holt that would otherwise never be approached and thus might increase disturbance. The two general approaches to reducing risk of theft are camouflage (seen in Figure 5.2b) and secure placement. Camouflage doesn't have to be natural colours or leaf netting, although that might be useful. In an area with litter, flood debris or infrastructure paraphernalia a camera-trap emplacement could also be disguised within a human-made object that would be unlikely to attract attention, such as a pipe (example in Figure 5.4) or even some litter. It is possible to buy boxes for camera-traps that lock the device in while not obscuring the PIR sensor, camera or LED array. While these might deter an opportunist thief, we have found that if someone wants to steal the camera-trap they may well come back with bolt cutters or even just remove the whole emplacement in one piece.

Theft and damage are another reason for swapping camera-traps every three weeks or so: if cameras are damaged or stolen you only lose data of that duration.

5.4 How should I deploy my camera-traps at potential couches?

Otters often rest above ground in couches. There are additional logistical considerations when camera-trapping couches:

- It is certain that the surveyor will cause disturbance to an otter occupying the couch when the camera-trap is set up and visited to change batteries and so on.

- A camera-trap that overlooks a sleeping otter for several hours will repeatedly trigger on movements the otter makes and so battery depletion is likely.

One solution is to set a camera-trap a few metres down the run to the sleeping area to pick up details of the otter and activity patterns. This sounds fine: no continual triggering on a sleeping otter, and further away so potentially less disturbance. However, if you are not monitoring the sleeping area itself, the otter could walk past the area that you think represents the sleeping site because the actual place of rest is 30 m inland, and the spot being monitored is actually a spraint site or a rolling pitch.

A better approach is to have a camera-trap set on time-lapse on the suspected sleeping area and programmed to take a still image every fifteen minutes. The 96 images you will have from each day will mean that the camera-trap can be left for a very long time without depleting battery or memory availability, so minimising maintenance visits to every two months or so. As the time-lapse camera-trap would not have to be triggered by the otter, it could be placed above the otter's eyeline/scent line, looking downwards. It could also be further away than would be recommended for good detection with a PIR, enabling a good field of view encompassing the whole of the suspected resting area. The addition of a camera-trap set to PIR by the run would add an additional layer of data and ideally would be better placed to record the sex of the otter. This camera would capture the otter's passes to and from your place of interest. If regular passes of otter were seen on the run camera, but no rests were observed on the time-lapse camera, it would suggest that either the place of interest was a spraint or rolling site or that the otter was simply passing through.

5.5 What is the best pattern and timing to camera-trap a resting place?

5.5.1 When to camera-trap holts

It is unlikely that there is an equal chance of a holt being used for resting and/or breeding throughout the year. Enclosed holts provide cover, shelter and insulation, which may be more sought after in winter and spring. In temperate climates such as the UK and large parts of Europe, there is more natural cover from vegetation and less need for insulation and shelter in the summer. In a radio tracking study in Austria, 60% of identified resting places (locations where an otter was detected to be passive for a period during the day) were enclosed structures, that is, holts (Weinberger 2016). However, use of such structures was more common outside the 'vegetation period'. In our camera-trapping study in southern Scotland and northern England, at six repeatedly used resting places, 95% of rests occurred in winter and spring (December to May), and only 5% in summer or autumn. Additionally, our long-term study holt in central Scotland had a bias of rests towards winter and spring (unpublished data). This may be because some structures have greater reuse from winter to spring because they provide better concealment and/or thermal properties, and/or because these qualities may make some structures stand out more as potential resting sites to a surveyor and are therefore more likely to be recognised during a walk-over survey.

As otters do not have a physiological mechanism that dictates when they breed, it is not possible to schedule camera-trapping to determine if a resting place is used by a female with cubs. Otters can breed in all seasons (Heggberget and Christensen 1994) and females can be in oestrus in any season (Hauer et al. 2002). While records from different European countries do indicate some seasonality of breeding, there is variation (Chanin 2003b). For instance, in Germany, the season with the fewest oestrus females (assessed based on road killed females) was summer, although samples size was low (n = 17) (Hauer et al. 2002). In the UK, a bias towards winter for inland Scotland based on unpublished data is also noted in Chanin (2003b) and it is fairly well documented that in Shetland breeding is strongly biased towards the summer (Kruuk and Moorhouse 1990).

We have gathered images/footage of young otters, records of otter cub rescues, and roadkill data provided by Cardiff University in the UK, and from initial analysis there seems to be a general bias away from summer births throughout much of the UK (Oliver 2023). However, this is a preliminary finding and may not reflect all areas. There are likely to be some complex patterns underlying any variation, such as patterns of availability of the main prey items in different ecosystems.

Based on these patterns, to detect resting and whether a resting place is used by females with small cubs, it seems likely that camera-trapping of resting places should at least include a winter (December–February) and/or spring (March–May) survey, and that a summer (June–August) survey alone is unlikely to be adequate.

5.5.2 Camera-trap duration to detect resting places

If the purpose of camera-trapping a potential holt is to identify a resting or breeding place, as soon as evidence of resting or breeding is observed, which we discuss in Chapter 6, the survey has achieved its purpose. The more challenging consideration is when to be sufficiently confident that the site is unused and therefore to stop camera-trapping. Otters usually have long, linear home ranges, so they can't exploit all the resources from a single central resting place. At the resting places we have monitored for long periods we have found that resting can occur regularly for several weeks, or erratically – resting can be detected once or a handful of times in short succession, but then no resting is detected for some time (Findlay et al. 2023). This unpredictability makes it challenging to know whether a period with no resting detected means the site is not a resting place.

To deal with this and address the minimum duration of camera-trapping required to have a 95% chance of detecting at least one rest at a known, habitually used resting place, we undertook simulations on resting data from known resting places to determine an optimum camera-trapping duration and strategy (through one or two periods of monitoring) (Findlay et al. 2023). Because 95% of our rests were from winter to spring (December–May inclusive) we excluded summer and autumn. We found that in our population it was more efficient to camera-trap for about 40 days in winter and repeat this in spring (about 80 days in total). For a single camera-trapping period, a greater duration was needed (about 100 days in total). Having two shorter periods of camera-trapping, one in each season, may work better than a single longer period because if resting occurs only in one of those seasons, the former reduces the risk of missing it. Because some of our Tweed sites were also used as maternal resting places, the same camera-trapping duration is recommended for maternal resting places as other resting places. A single record showing resting by a female with cubs less than six months old would be sufficient to confirm a maternal resting site.

From discussion with other ecologists this may be surprising, since we have known UK ecologists to have camera-trapped potential resting places for much shorter durations than this, and during the seasons when we saw little or no resting in enclosed structures. Although the suggested duration (at least two periods of 40 days) and timing (in winter and spring) we found in the Tweed catchment may not be representative of all regions, we would still recommend camera-trap surveys of substantial duration (five to six weeks) be conducted in more than one season (ideally including winter and/or spring) to confidently rule out a site as a resting place if no resting activity is detected.

5.5.3 Camera-trap duration to detect natal holts

A natal holt can only be identified when it is functioning as a natal holt. However, a camera-trapping duration of six continuous months may well yield no natal or breeding activity. This is insufficient to state that the structure is not a natal holt because it will only have this function for a limited period.

If a structure is thought to have potential as a natal holt a different approach is needed to that we described for resting places, because the natal holt functions as such for eight to ten continuous weeks per litter and, most of the time, it would be expected that the holt has a maximum number of one litter per year. To avoid causing disturbance during the sensitive natal period, visits to change batteries and SD card should be minimised and maintenance visits should aim to be approximately every six weeks, or longer if it is thought that the batteries and SD card will last. Breeding can be at any time of year, therefore a full year of camera-trapping would be necessary. This may sound excessive but it means only eight or nine very brief site visits. The data collected does not have to be exhaustively searched for natal activity, which would entail days of watching videos of false triggers, mice and birds with the occasional otter. Instead, the data can be sampled so that one week's worth of footage is watched in each month. If a resident female is noted, additional days can be viewed until it is determined whether or not the holt is natal (i.e. small cubs are seen emerging). This minimises visits to the holt as much as possible and reduces video viewing time.

The above sampling schedule assumes that females do not share natal holts and that a maximum of one litter is expected per female core range. The core range is the area of the female's home range where she gives birth and rears the young cubs, which is generally smaller than her complete home range. Accepted wisdom on breeding also suggests that females' core ranges do not overlap. However, we have recorded the full natal period of three females giving birth in our long-term study holt in one autumn to winter period; not enough is known about breeding ecology to say whether this is exceptionally rare or a regular occurrence.

5.6 Other applications of camera-traps for otters

We have discussed in detail best practice for camera-trapping potential resting or breeding places, but there may be other scenarios in which camera-trapping away from resting places is useful or necessary, which we discuss below.

5.6.1 Assessing age of emerged cubs

Females with young of up to about six months old are more vulnerable to disturbance. If a female and young cubs are accidentally flushed from a maternal resting site and become separated, there is no guarantee that they will reunite or that the cubs will survive on their own. Some construction or management operations may need to be timed to align with the breeding status of the female, proceeding only when she has older cubs that are more robust. In areas with a known or suspected female and cubs, camera-trapping at spraint sites or along runs may provide footage of the female and any accompanying young. You can then gauge whether the cubs are very small or older (advice on ageing cubs is in Chapter 6), or whether you have an adult female with no accompanying cubs, in which case she may be pregnant or have cubs in a natal holt. You can then extrapolate forwards to a time when any cubs are robust, which would be the optimal window to undertake operations. You may need to camera-trap several spraint sites to gain this insight, because you cannot assume that the female and cubs always spraint at the same spraint sites, although they often do. Note that the gestation time for otter is approximately 63 days and that the newly born cubs stay within the natal holt for 8 to 10 weeks.

5.6.2 Spraint sites

Following most of the principles laid out for camera-trapping potential resting places is a good start if you plan to camera-trap at a spraint site. Importantly, no licence is necessary to camera-trap at a spraint site or along an otter path unless you are close to a resting site. Trapping duration is far more bespoke for this, and you can get enough data within three to four weeks. Likewise, the time of year is not critical, although as otters do not maintain spraint sites so rigorously in summer it may be best to camera-trap in other seasons. You will need to sex the animals if your aim is to identify potential breeding females, making it even more important to set your camera-trap at a low height. If you have a steep bank the camera-trap could be set lower down than the spraint site to get a good view of the rear workings of the otter (Figure 5.5). You would also see nipples if the female was lactating. The video setting would be essential to assess where the urine stream emanates from, which is discussed in the next chapter.

5.6.3 Paths

There may be instances during which you need to camera-trap an otter path, for instance if you need to assess its 'frequency of use'. Thought is needed before embarking on such an exercise as to how informative any data would be. The angle of camera-traps to the path is important because the otters may be moving quickly. Camera-traps set perpendicular to a path are not recommended: if they

Figure 5.5 Setting your camera-trap at a low height enables a better view of characteristics needed to sex an otter, which is important if you are trying to establish whether you have a breeding female. This otter is a male! (Barbara MacFarlane)

are close enough to be triggered, the passing animal can often be missed because it is already past the field of view by the time the camera-trap takes an image/video; if it is set back, the camera-traps have a lower trigger probability. The best approach is to set camera-traps close to and at a tight angle to a path so that they are looking broadly along it. This needs to account for topography there should be no humps or bumps between the camera and the path and the angle of the camera in relation to the sun at dawn or dusk. A diagonal camera-trap will have some directional bias because capture will tend to be better when the animal is moving away from the camera-trap, since it is likely to be within the field of view for longer after triggering (Apps and McNutt 2018). For this reason, having two cameras pointing diagonally in different directions along the path is optimal for capture in both directions, albeit placed so that they don't face each other and cause glare if both are triggered at night.

As at spraint sites, setting camera-traps at a low height along paths may increase your chance of sexing the otters, as would setting them on any rolling pitches or spraint sites where different angles of the otter could be obtained. The best height to set the camera-trap is at the chest height of the target species. However, otters are very low to the ground and vegetation between the camera-trap and the path may obscure both the field of view and the detection zone. In such cases the camera-trap should be set slightly higher and angled slightly downwards to compensate.

5.6.4 Monitoring otters during construction

There is often a requirement for an ecological clerk of works to monitor otter activity during a construction project. It is important to realise that while lots of spraint indicates frequent presence of otter, lack of spraint does not indicate their absence nor even a lessening of otter activity. Likewise, camera-traps could be set up to record otter activity, but a camera-trap is effective over a very small area and so lack of otter registrations does not mean that otters are absent. Monitoring otter activity using spraints or camera-traps must therefore be thought about in terms of what is actually possible.

5.6.5 Monitoring mitigation

Common mitigation measures include provision of artificial holts and facilities for otters to negotiate roads, such as dry culverts or retrofitting otter ledges into culverts above the level of the water or where the watercourse is on a steep gradient and has been 'stepped' too steeply for otters to travel upstream. Both of these can be monitored using camera-traps.

Artificial holts

Monitoring artificial holts should follow the same protocol as for camera-trapping natural resting places. However, it is possible to design the artificial holt to include an entrance lobby within the structure that is not the actual resting chamber. Such a lobby is ideal for camera-trap emplacement and should be easy to reach but out of sight. And because it is inside the structure it avoids the false triggers caused by light and waving vegetation. The only drawback is that mice also like such structures and may party each night directly in front of the camera-trap. Nevertheless, such emplacements can last for months without a battery change. It is advisable to camera-trap the artificial holt entrance from outside the holt for a couple of nights prior to retrieving the interior camera-trap, to avoid having to put your hand into a structure that an otter may be occupying, which would disturb any otter using the holt for resting.

Culverts and ledges

Monitoring dry culverts, ledges or other structures below roads is more difficult unless the aim is just to gather evidence that they have been used by otter (such as where the ledge provides an access along a stepped watercourse). Evidence that the mitigation has worked would require one or more otters to be passing through the feature during spate conditions.

With this in mind, a photograph of an otter at one end of a linear tunnel is not undisputable evidence of use. Ideally, a camera-trap at each end of the tunnel would be needed to show that the otter has actually gone through and hasn't done a U-turn part way. The field of view of the camera-trap needs to be carefully considered in such cases and also the detection probability. For instance, if the culvert contains water, an otter could swim through undetected or could walk past the camera-trap because it is wet and has lower detection probability. The survey duration and seasonal timing needs, therefore, to be carefully considered. In addition, either long-term monitoring is needed so that enough spate conditions are encountered to consider that passage of an otter is likely, or reactive monitoring involving placing camera-traps just before spates are predicted. It is therefore quite an easy sentence to include in a species protection or monitoring plan, but expectations need to be realistic.

5.7 What to do prior to set-up and during set-up, maintenance visits and take-down

5.7.1 Tests and storage before you set up

Camera-trap models vary in the extent to which you can be confident they are functioning as required when left in place. Many models will have some kind of indicator that they are armed or arming when you turn them on (such as a flashing light, or if they have a front-facing display they may have a countdown timer on screen). Outside of this, it is good practice to do some general tests both on the model of camera-trap you are using, so that you understand its function well, and also on each unit of that model you use, to check it is working properly.

The trigger time of a camera-trap will often be advertised on the box, and this is likely to vary between video or still images as discussed. If you don't know the trigger time for your camera-trap for the type of recording intended (video or still images), or you suspect it is not as advertised, it is possible to test the trigger time yourself. One approach is to place a digital stopwatch that displays tenths of a second, such as on your smart phone, facing the camera-trap so that it is in the field of view of the camera (and in focus). If you hold your finger still over the start button for the stopwatch, then press and pull it away quickly in one movement, the value of the stopwatch seen in the first photo, or the first frame of the video, should give an estimate of the trigger time. As there is room for error in this, it is worth doing several times and working out an average – it only needs to be approximate, enough to compare between models, modes or to detect any overly long delay.

Another test you can do before going into the field is to check that all the IR LEDs are functioning and whether they emit a red glow. As noted, we strongly advise against using low-glow (red IR) camera-traps at holts and, indeed, your licence conditions may stipulate that they cannot be used. Nevertheless, you should test your camera-traps in the dark to check that there is no red glow (even if the model states that they are "no glow") and that the LEDs are functioning. Note that for some cameras, the lower illumination settings will mean that some LEDs are kept off, so testing should be done on all the illumination settings if applicable. Even if your camera is a no-glow model, as we recommend, you can still typically view the LED flash by viewing through a smart phone camera – these are usually sensitive to IR LEDs and you should be able to see them on the phone screen. If the room is also silent, this is a good opportunity to check that the camera is not emitting a noise when it triggers, such as a whine or click.

Your camera-trap will have its own smell and is likely to carry the scent of the person(s) who handled and placed it, as is the immediate area around it, although this is likely to diminish over time. We advise where feasible that camera-traps are kept outdoors for a period before deployment, to diminish human smell, and also that they are kept away from dogs.

It is strongly recommended that you pre-programme your camera-traps with the desired settings before placing them out. To understand any otter activity patterns (discussed in the next chapter) it is essential that they are showing the correct time and date, and that two camera-traps at the same holt are synchronised as precisely as possible. Camera-traps can lose their date/time settings if the batteries are taken out, so if you are redeploying the same camera-trap you need to check that the programming has retained the correct time/date and not reverted to a default. If you have two camera-traps monitoring the same area, after taking the camera-traps in, it is worth checking that their times have remained synchronised. Find a unique video of an event recorded on both camera-traps near the end of the trapping period, for instance a blackbird with a worm or your hand waving before you take in the camera-trap, and check that the date and time of the two camera-traps match.

5.7.2 Set-up

Setting up is likely to be the longest time you spend at a potential resting place, so good planning is required to minimise it. Ideally, camera-traps should be set up by as few people as possible (one or two) and with minimal noise. Prior planning about camera-trap emplacements is essential so that the camera-traps can be installed as efficiently as possible. Setting up should avoid the period around sunrise and sunset. Checking that the camera-traps have the field of view you planned is important, as it can be easy to think that they are pointing in the right direction only to find later that they are not quite right. Some models have a handy viewing screen that allows you to see what the camera-trap is seeing when in position, but others have no screen or a screen that is only viewable when the camera is hinged open. For the latter models, a useful piece of kit is an SD card reader that can be plugged into a smart phone or tablet. Using this, you can place a camera-trap, trigger it and quickly check the footage while in the field.

Finally, we recommend that each camera-trap is deliberately triggered by waving an arm in front of it after you have set up and again when you swap or collect it. This is a useful check to see if the camera-trap has functioned through the whole period, especially if you had several days prior to the end of the period where there is no footage at all. If you see your waving hand at the start and the end of the deployment, you know the trap was functioning throughout. It also confirms whether your camera-traps were properly synchronised, if you used more than one on the same holt. If your waving hand is absent at the end of the deployment, you must assume that the last trigger (whether of an otter, mouse or vegetation) represents the last date/time that your camera-trap was functioning. Note that this technique will not work if you have set your camera-traps to only record during nocturnal periods, because setting up and maintenance visits should only be done during the day.

5.7.3 Maintenance visits

'Maintenance visits' refer to visits to camera-traps between set-up and take-down to change SD cards and batteries and check that the camera-traps are functioning and still in place, or simply to replace the camera-traps with fresh ones. Because we recommend relatively long surveys of several weeks, most deployments at holts will require one or more maintenance visits. We have found that maintenance visits of about once every three weeks strikes a good balance between (i) the risk of data loss if SD cards get full, batteries are depleted or cameras are damaged or stolen, and (ii) minimising visitation frequency to avoid unnecessary disturbance. Longer periods between maintenance visits should be considered if monitoring a suspected natal holt, and four to five weeks should be considered between surveyor visits. Longevity of the battery and SD card could be increased by setting duty time, to avoid the four hours or so in the middle of the day and to reduce false triggers. Some registrations of otter may be missed, but over a period of six weeks this would be unlikely to affect recognition of activity patterns of a female with small cubs in the holt.

At resting places, we recommend swapping out camera-traps at maintenance visits. This means taking a pre-set camera with the desired settings and simply swapping the fresh camera-trap for the one that has been in the field. This allows you to spend far less time at the holt compared to fiddling about to change batteries and SD cards. Time at the emplacement can be reduced to 20 seconds per camera-trap if swapping in and out of fixed emplacements, compared to the several minutes that can be required to change camera-traps fixed by cumbersome straps (which also make it hard to match the same field of view each time) and replacing batteries and SD cards. The camera-trap you remove can be taken back to the office/home/vehicle, checked for any problems, wiped over and refreshed with batteries and SD card ready to be redeployed later.

Swapping out does mean you need more camera-traps than are deployed in the field at any one time, although, if planned carefully, you can take the swapped-out camera from one site, check and replenish its batteries and SD card, and then swap it with a used camera-trap on another site (and so on, in a chain), so you may not need double the number of camera-traps. Either way, if you are camera-trapping commercially, you should be able to justify this need for more camera-traps in terms of best practice, so that the cost of any additional camera-traps can be borne by the client. The major advantage of swapping in a fresh camera-trap is that if you have a faulty camera-trap, then it only remains in the field for one monitoring session. The worst thing is to swap SD cards then get beck to the office and find that the camera-trap that you have left out is not working.

Although generally we recommend swapping out whole camera-traps during maintenance visits, where this is not possible and you need to change batteries and/or SD cards in the field, water from the outside surface of the camera-trap, or rain, can easily get into the camera-trap. A tip we were given by a student of ours is to carry an absorbent microfibre cloth and dry the outside of the camera-trap as far as possible before opening it. This can also come in handy for cleaning any smudges or muddy badger snuffles on the Fresnel lens.

5.7.4 Data management

The organisation of data should be planned before any but the simplest of camera-trapping monitoring. It is so easy to get sites mixed up, or store files in places and forget where they are. Labelling SD cards and naming camera-traps for use on specific sites can help, as can having a well-structured filing system. Batches of images or videos can be renamed using renaming software too, which helps retrieval. File names could include the site and camera-trap position followed by the date and time of the capture. For example Thames_S4_CT3_251004_22:12:37. The renaming software usually gives the choice to use the metadata of the file (the date and time) as part of the file name so all you need to specify is the prefix of the site and camera-trap position. You then choose the date and time format.

5.7.5 Checklist for what to do prior to set-up and during set-up, and maintenance and take-down visits

We have compiled a checklist for preparation, set-up, maintenance visits and take-down that we recommend using, at least initially (Table 5.2). Some aspects relate to things discussed in Chapter 6 but are here so that they are together.

5.7.6 Reporting details of method

When producing a report it is crucial to include the details of the camera-trap set-up and programming. Some of this information is very standard and could be included in a table in an appendix (an adaptation of Table 5.3, for example), with the main text summarising the rationale such as the placement of the camera-traps.

5.8 Problematic sites

Not all potential resting places will be easy to camera-trap. We have already discussed using bespoke emplacements to deal with complex topography, and here we discuss even more challenging scenarios.

5.8.1 Entrances close to water

A structure that has an entrance just above water is problematic in several ways. Firstly, your camera-trap may be prone to submersion if placed low, as recommended, and secondly, the detection probability of the otter during the brief time it takes to come out of the water and go into the entrance will be low, because a wet otter's surface temperature can be very similar to that of the background. Having a really close camera-trap, within 1 m, will improve the detection probability, but you need to satisfy yourself that this won't impede the otter's usual routes. At such a close distance the otter is likely to be very large in your field of view, which is not ideal. This can be considerably helped by using a camera-trap that has a wide-angled lens, which makes the otter look further away and smaller (see Figure 4.2). Camera-traps

Table 5.2 Checklist for what to do before you set up and during set-up, maintenance and take-down visits. More information on each is provided in the text. This list can be scanned and printed for use in the field.

Phase or visit	Essential	Tick when done
Prior to set-up	Ensure you have the correct licence and that you understand its conditions.	
	Ensure you have the landowner's permission to camera-trap.	
	Ensure you have adequate file storage space and back-up capability for the footage collected, which can be substantial.	
	Consider camera-trap placement and ensure you have the correct materials for any planned emplacement.	
	Ensure camera-traps have the correct date and time and, if more than one is used, are synchronised.	
	Check that camera-traps have the correct settings (e.g. videos, 20 seconds duration, minimum delay between triggers, low IR illumination).	
	Check camera-traps to ensure they are working.	
	Charge batteries; check SD cards are empty and working and have been reformatted in the camera.	
	Store camera-traps to minimise human or pet scent.	
Set-up	Task just one or two people with set-up, with no dogs present and phones on silent.	
	Use an auger to prepare a hole for the stake and firm it in. Do not bang in stakes.	
	Set camera-traps at an appropriate distance (1–3m) and height (20–30 cm) and ensure they are level with the area of interest.	
	Identify any otter paths; avoid them with camera emplacement.	
	Carry out during daylight hours.	
	Check fields of view of camera-traps.	
	Deliberately trigger camera-traps before departure.	
Maintenance visit	Deliberately trigger camera-traps on arrival.	
	Swap all camera-traps with replacement traps containing fresh SD cards and batteries, or replace all batteries and SD cards on placed camera-traps with fresh ones.	
	Carry out during daylight hours.	
	Allow only one person at the emplacement, no dogs, phone on silent.	
	Deliberately trigger camera-traps before departure.	
Take-down	Deliberately trigger camera-traps on arrival.	
	Remove all equipment (unless emplacements are to be used in a follow-up survey).	

Table 5.3 Useful list for reporting set-up and programme settings

Category	Note
Equipment details	Make and model of camera-trap. Trigger time if known, either still or video. Illumination frequency (black IR etc.).
Programming	Video or still. Details of video or still (e.g. 3 stills, 1 second apart, 20-second video). Trigger delay, state choice of programme. Illumination setting, state choice of programme. Duty time, state the hours during which the camera-trap is functioning each day.
Set-up (an overview photograph or a diagram is very informative)	Number of camera-traps on target area. State height to PIR sensor and explain if not at chest height of target species due to terrain or vegetation. State distance between target area and camera-trap (guesstimate in preference to leaving scent on target area and taking time over using a tape measure). Distance to water (this affects detection probability).

placed close to the water in rivers prone to spates may need to be collected before forecasted spates and then redeployed.

5.8.2 Underwater entrances

There is no currently available technology that caters for resting places with underwater entrances. If you have a holt with underwater and terrestrial entrances, the latter should be camera-trapped. In such cases you may have numerous videos of otters coming out of the structure but none going in, because they may be entering the structure under water. It is still worth camera-trapping the terrestrial entrances because behaviours and activity patterns can indicate resting even if you have incomplete data, as discussed in Chapter 6. However, if you suspect a potential resting place that only has underwater entrances, camera-traps obviously are not an option and you will have to treat the structure as a resting place under the precautionary principle.

5.8.3 Complex habitat

The ideal scenario for camera-trapping is to have a clearly defined place of rest with a single entrance so you can readily monitor how long an animal remains inside. However, there are often cases where you suspect a resting place in an area of dense scrub, a chaotic area of boulders such as you find on the west coast of Scotland, or a run of rock armour. In these cases you still have a chance of detecting resting if you monitor paths radiating from the feature and look for patterns of 'ins' and 'outs' (discussed in Chapter 6). The point at which an otter path enters the complex feature may not necessarily be where the resting chamber is, as there could be many hidden passages between there and the actual place of rest. Nevertheless, activity patterns observed may be enough to confirm (or confidently assume) resting, even if you are not able to pin down the exact spot. In these cases, you will have to consider a larger area/feature as protected.

5.8.4 Reedbeds

Setting up a camera-trap in a reedbed is fraught with issues:

- Stomping through a reedbed to the camera-trap site will disturb any resting otters (and other wildlife) over a considerable distance.
- Rising water may prevent access on foot to the camera-trap for maintenance visits and may also flood the camera-trap.
- If a stake has been used and moves due to wind or water, the camera-trap is likely to repeatedly trigger.
- Moving reeds in the wind are likely to cause excessive triggering.

If the structure of interest is considered to have been constructed by an otter, such as an 'igloo' (e.g. Figure 3.21), then it can be confirmed as a resting place without camera-trapping. If the objective is to monitor an open structure, taking time-lapse footage, as per section 5.4, will overcome some of these issues, and the frequency of maintenance visits required could be significantly reduced. Setting the camera-trap at a greater height may be beneficial to reduce the chance of submersion. Ultimately, the risk of disturbance has to be balanced with how critical the requirement for data is; monitoring within the reedbed should only be undertaken if absolutely essential and only if the data gained is predicted to be robust.

5.9 Dealing with faults during camera-trap surveys

5.9.1 Camera-trap fault, environmental conditions or faulty battery?

If you find that a camera-trap appears to have malfunctioned, a first recommended check is whether the batteries have any power. If they are flat, you should check whether you have an excess of false triggers (e.g. over 1,000), in which case this may be the reason for battery depletion. This could be due to a fault with the PIR, or deployment where funky light or waving vegetation causes excess triggers. There are three possibilities if the batteries are flat in combination with an unremarkable number of triggers: the batteries were not fully charged when you set up, one of the batteries was faulty or the camera-trap was faulty and the batteries discharged. If you are using rechargeable batteries, recharge the batteries and use a handheld voltmeter to test each battery to see if one carries a lower charge than the others. If so, you should discard the faulty battery, reload with a set of fully charged batteries and test your camera-trap by running it in the office, house or garden for a few days to be confident that it is running well. Other malfunctions such as water ingress or a corrupted SD card are possible. Camera-traps that have had water ingress should be thoroughly wiped, then dried out as soon as possible by placing in an airing cupboard or other warm area for a few days. This may well bring them back to life and can also defog internal viewing screens. This is another advantage of swapping camera-traps, because it means starting each period between maintenance visits with a guaranteed functioning camera-trap and that you are not placing fresh batteries and SD card into a broken camera-trap without realising it.

Some faults with camera-traps will be unfathomable, but if a camera-trap fails it is worth checking that all the batteries are making good contact and that there aren't any dry joints (failed soldering). We have had camera-traps fail when the IR has 'stuck on' and so continuously illuminated and drained the batteries. We have also had camera-traps fail when the IR illumination flashes on and off constantly, independently of the camera coming on, again draining the batteries. These faults were not obvious and we only found them when reviewing footage of the camera-traps on CCTV, when the camera-traps could be seen to be misbehaving.

5.9.2 Keeping some back-up camera-traps

Don't assume all your camera-traps will work all the time. As well as needing the same number of camera-traps as there are emplacements for a study, plus additional ones for swapping out, some additional spares will be extremely useful. Your study could come to a halt or become less effective if camera loss or malfunction leaves you short.

5.9.3 SD card cleaning

Occasionally you may find an SD card is not registering in a camera-trap, which can be due to a poor connection, for example if there are fingerprints or dust on the terminals. One trick that can work is to give the contacts on the SD card a gentle wipe with a clean pencil eraser, which can make them function again.

5.10 Key points

- Camera-trapping can be essential to reassign a potential resting place as a confirmed resting place and determine if it is a breeding place.
- It is critical that no-glow camera-traps are used and that a code of conduct is followed to minimise potential disturbance.
- For enclosed structures, where the camera-trap is not monitoring the actual place of rest, we recommend setting your camera-traps to record video clips (not still images) of about 20 seconds, with no delay set between clips.
- For unenclosed structures where the sleeping otter would be in the camera-traps detection zone and field of view, we recommend time-lapse images.
- Where possible, using two camera-traps per potential resting place is important for several reasons, including insurance against one camera-trap

malfunctioning and to build up a more complete picture of activity and behaviour.

- At potential resting places, camera-traps should be set at a close distance (1–3 m) to maximise detection probability and, where possible, at a low height of 20–30 cm to aid identification of sex.
- Careful planning and use of predesigned emplacements that allow camera-traps to be swapped out quickly (within a few seconds) every three weeks or so are likely to substantially reduce time spent at the site, reducing potential disturbance.
- Use of enclosed resting places (holts) may have a seasonal bias, thought to be due to a greater requirement for concealment and insulation in winter and spring. If a summer or autumn camera-trapping survey is undertaken and no resting is encountered, additional camera-trapping should be undertaken in the winter to spring period before a structure can be discounted as a resting place.
- To identify whether a site is a resting place or not, we recommend at least 40 days' (about 5–6 weeks) camera-trapping in winter and 40 days in spring. If a rest is detected, even once, camera-trapping can cease because that confirms it as a resting place. We recommend the same trapping duration to identify maternal resting places. Maintenance visits are recommended every three weeks or so.
- To identify natal resting places, we recommend camera-trapping the holt for a full year, with as few visits as possible to maintain the camera-traps. The resulting video data should be sampled for efficiency, with one week of footage per month being viewed to identify daily occupation by a female.
- In some scenarios it may also be beneficial to camera-trap spraint sites or otter paths, and these have their own specific recommendations.
- Careful checking of camera-traps before deployment and after collection is crucial for a smooth and effective camera-trapping survey.

6. Interpreting camera-trap data from potential resting places and other sites

6.1 Introduction

This chapter is a practical guide to help you turn your camera-trapping survey into data that can be accurately interpreted.

We have established that spraint at a structure isn't evidence of resting, and when it comes to camera-trapping, an otter observed at a structure, or even entering a burrow, does not indicate resting. Taking this a step further, otters observed at a structure on 'many' nights also infers nothing over and above the simple fact that otters were observed. Simple statements made from a quick look through your videos are poor summaries of good data, which, if recorded and examined in an analytical way, may actually show more than 'otters observed on many nights'. This chapter will help you take that crucial step from simply thumbing through otter footage and reporting in qualitative terms, to methodically documenting your otter activity and key behaviours from your footage so that you can summarise the data and interpret it for impact statements and, if appropriate, licence applications.

Some sites may yield very few registrations of otter, and so the stages described will be very quick. Other sites can be extremely busy, with near-nightly registrations of otter. At the time of writing we have a site where over 900 20-second videos were taken in a three-week period, of which 200 were of otter. Just skimming through the footage to pick out rests (including natal activity) was not possible.

It is useful to think about the information gleaned from the footage being arranged in layers, as in Table 6.1. With a full dataset, these complement each other to give a full account, but some aspects of these layers can provide standalone evidence either of resting, breeding or natal use of a structure.

The following chapter introduces a method to prepare, organise and record your data as an 'events list'. An events list is a spreadsheet that can be thought of as being like a diary of events, or a timeline of the otters that interacted with the structure. This timeline is the backbone of the events list, but additional layers can be added, such

Table 6.1 Layers of information from camera-trap footage.

Layer 1: activity patterns	Layer 2: otter information	Layer 3: behaviour
• Duration that otters are inside the holt or occupying a couch • Date • Time relative to sunrise and sunset • Season	• Sex • Age • Breeding status • Individual otter or family group	• Use of latrine • Bedding collection

as information on the otter(s) seen and selected behaviours, according to the needs of your project. We describe a tried-and-tested workflow to constructing an events list and then give guidance on how to interpret the activity and behaviour in terms of discerning resting, breeding and natal activity.

6.2 Survey effort

Before starting to look at each otter video it is a good idea to understand how the camera-traps have performed. There is no point in stating that camera-traps were deployed for four weeks if the SD card filled up on day 14, meaning that the second half of the monitoring period was defunct. Document the first date of each monitoring period and the date the camera-traps were taken in or when the batteries were changed. If you followed the advice in section 5.6.2 and triggered the camera-trap when you set it up and again on retrieval, images of your hand on the first and last image will be a strong indication that your camera-trap has worked throughout the monitoring period. If you didn't do this (or couldn't due to duty time settings), record the date when the last footage was taken. If this is the same date as that when the camera-traps were removed, it's reasonable to assume that the cameras were functioning throughout the monitoring period. However, if the SD card filled or the batteries ran out of juice prior to that, the last date the camera-trap was functioning should be stated. The number of days each camera was recording can then be calculated and reported.

This information is easily documented on a spreadsheet (Figure 6.1) which is separate (i.e. on a different tab) to the events list. The 'Potential CT days' (potential camera-trap days) can be calculated by subtracting the 'Start' column (column B) from the date

	A	B	C	D	E	F	G
1	Cam	Start	End	Last Footage	Potential CT days	Actual CT days	Reasons for reduced recording
2	A	01/10/2024	30/10/2024	30/10/2024	29	29	NA
3	B	01/10/2024	30/10/2024	10/10/2024	29	9	Battery depletion

Figure 6.1 Example Excel spreadsheet recording camera-trap survey effort to show dates deployed and any periods when the camera-trap was not functioning. NA denotes 'not available'.

and time in the 'End' column (column C) by inserting the formula =C2-B2 into cell F2 and pressing the return key. Likewise, the 'Actual CT days' can be calculated by subtracting the 'Start' column from the 'Last Footage' column. Before calculating with dates, make sure that the format of the cell or column is set to 'Date' format.

6.3 Preparing the footage

6.3.1 Separating captures of otter from other footage

This is an optional step, especially if there isn't much otter activity. However, it can be very helpful if there are numerous videos of otter plus a mix of false triggers (e.g. from waving vegetation) and other non-target species such as the inevitable volume of mice dashing around. The idea is to set up a group of subfolders for each camera placement so that the footage you are interested in can be separated out. Three subfolders work well if you are interested in the presence of other mammals: (1) 'Otters', (2) 'Other mammals' (i.e. mesocarnivores), and (3) 'Trash' (any footage not in 1 or 2). This can readily be reduced to two subfolders: (1) 'Otters' and (2) 'Trash' by treating the other mammals as Trash if you are not interested in them.

A scan through the thumbnails is a good place to start, cutting and pasting all obvious otter footage into the 'Otters' folder. It is worth remembering that the thumbnail is not necessarily the first frame of the video, and so thumbnails that are otter-less can contain really good otter passes, so don't dismiss that footage straight away. For instance, if two otters run past, the first may trigger the camera-trap but not get recorded, and the second otter may appear after ten seconds or more and so is not on the thumbnail. It is therefore important to realise that video thumbnails are *very* unreliable representations of whether the camera-trap video contains an otter or not. Badger and other mustelids, and foxes, are similarly sorted into the Other mammals subfolder, or Trash according to your needs.

Now comes the tedious task of going through the remaining videos. Many will be rodents, birds and triggers caused by waving vegetation and can be put into the Trash subfolder, but there may well be occasional videos with an otter's tail on, or even just vocalisation, which should all be added to the Otters subfolder. While your laptop's memory may be begging for mercy, it is very important that you don't discard the Trash folder just yet, as some of the footage may be important. This is discussed in section 6.4.1 (*Single event*).

Repeat the above for each camera-trap, even if you have a dual set-up (i.e. two camera-traps monitoring the same feature).

Use of AI

It is worth noting that even if in a few years all ecologists are able to use AI to go through their camera-trap video footage and reliably pull out videos with otter, there will always be a need for interpretation by a competent ecologist with an understanding of the limitations of camera-trap footage and knowledge of otter ecology and behaviour. Ecologists will still be needed to make many of the judgements discussed in this chapter, such as whether the camera-trap set-up was robust in the first place, whether we can pair up footage to form 'paired events' that may then represent rests, whether a glimpse of a tail flicking out of shot is an otter leaving the structure, whether the urination pattern of an otter indicates a male or female, and many others.

6.3.2 Ordering the footage

It is best to make sure that the video files are sorted by the time they were recorded, so they are watched in time order, the oldest to the most recent. This means that you can keep a mental track of whether any otters are in the structure, and for how long.

6.4 Layer 1: activity patterns

6.4.1 The events list

Regardless of how busy the structure or feature being monitored is, the primary aim is very straightforward: to create a timeline of when an otter is at the structure and how long it is there for. This is the events list. Simple sites, with few registrations, can be recorded and interpreted in exactly the same way as the more complex scenarios that follow; they will just be quicker. Likewise, open structures are very easy to document as the otter is in the field of view and so resting/sleeping is directly observed. Even so, summarising data as a timeline of rests is very informative. This timeline is best recorded on an Excel spreadsheet, with each continuous period of otter activity recorded as an observation (a row) so that each observation contains the date and time the otter starts interacting with the structure, and the date and time it leaves the structure, as per Figure 6.2. It is important with a closed structure, where you can't see the sleeping area, that the observation records whether the otter goes into the structure or just passes the entrance.

If you plan to do simple calculations in Excel it is essential that the columns are correctly formatted. For a column to accept a date and time you need to use a custom format set to dd/mm/yyyy hh:mm.

Paired events

The perfect record is a 'paired event'. By 'paired' we mean that we have captured a given otter's arrival at and departure from a site, regardless of how long it remains there. At its simplest, this could be an otter arriving at a burrow, going into the burrow, coming back out again and leaving, all on the same 20-second video clip. This would be represented as a single row of data with the date and time bookmarking the start and end points of the event (see Figure 6.2, observation 1). A slightly more complicated scenario would be the otter going into the burrow, triggering the camera-trap, and then the camera-trap not triggering again until two minutes later when the otter is recorded leaving the burrow (Figure 6.2, observation 2). At another extreme, two otters could be captured going into the burrow followed by 40 videos of waving vegetation, then 10 hours after the otters went in, the otters come out, loaf for a couple of minutes before leaving. There could be 8 or so 20-second videos recording this otter activity. This would be represented by two

	A	B	C	D	E
1	Explanatory		Layer 1: activity		
2	OBS. NUMBER	OBS. TYPE	OTTER IN	OTTER OUT	OTTER PASS
3	1	Paired	23/10/2024 02:30	23/10/2024 02:30	
4	2	Paired	23/10/2024 03:00	23/10/2024 03:02	
5	3	Paired	24/10/2024 07:30	24/10/2024 17:30	
6	4	Paired	24/10/2024 07:30	24/10/2024 17:30	
7	5	Paired			24/10/2024 17:00
8	6	Single	25/10/2024 07:00		
9	7	Single		27/10/2024 13:50	

Figure 6.2 Example Excel spreadsheet showing an 'events list' populated with activity, the first two columns in grey are explanatory.

rows of data (Figure 6.2, observations 3 & 4) – two events, one for each otter. Each event would start with the otter going in, and end when it leaves. Representing groups of otters as an equivalent number of events (i.e. one otter per row) allows for the animals to enter or exit at different times.

Otter *passes* should also be considered as paired events, as the entire activity of the otter between arriving and leaving the structure is recorded. Note that this can just include an otter sniffing and sprainting outside the structure, and leaving without going inside (Figure 6.2, observation 5).

Single event

A 'single event' is where either the entry into the holt or the exit has not been recorded (Figure 6.2, two examples – observations 6 & 7), and these provide less information, particularly if it is of interest whether the otter has rested or not. Various factors can cause single events, and they are especially likely if the camera-trap set-up is not ideal. The Trash folder can come in very useful when trying to resolve single events. If the otter was only recorded going in and not coming out, there is a good possibility of a false trigger immediately after the video of the otter going in. This occurs when the otter has exited in the split second during which the camera-trap is processing the last video and setting itself for the next. The camera-trap is triggered, but the otter is out of frame. Sometimes you can hear the otter vocalising as it moves away, or the rustle of vegetation. In such cases a single registration can become a paired registration if it's clearly a case of missed detection. In other cases you may simply not be able to find the single event's 'pair' and will just have to note it as such, although if you are getting lots of single events it could be worth considering adjusting your set-up.

Groups of otters

Groups of otter at a structure can get a bit complicated because they don't always enter and leave the holt at the same time. With young families, where the mother otter is clearly much larger, it can be useful to separate the mother's times of entry and exit from those of the cubs, especially as she may leave the cubs in the holt while she goes off alone. When the family is older and it is harder to tell the mother and cubs apart, a practical compromise is to record the time of the first otter going into the structure and the time that the first otter leaves to get a 'minimum' duration in the holt. Practically speaking, though, a family group of three otters is going to be difficult to miss, and providing evidence of this is the key, rather than a precise measurement of the duration of the rest.

How to compile an events list

Producing an events list is the ultimate goal in identifying patterns of activity and deciding whether you have a resting place, natal resting place or maternal resting place. With a single camera-trap, this can be very simple because all recordings are unique. With more than one camera-trap, either on the same entrance or on different entrances, this can be more complex. Some activity can be recorded on more than one camera, while other activity may only be recorded on one camera-trap. In the events list, it is important to avoid duplication, so if two camera-traps record an otter going into a structure, the events list must only include one entry. Often, jotting down activity at each camera-trap can help avoid duplicate observations, leaving a list of discrete observations that feed into the events list. This could also be achieved in Excel, by listing activity from each camera and deleting duplicate entries. However, this would then have to be converted into an events list, as a video-by-video account of activity is difficult to interpret.

A reasonable compromise is to view and record the events on the busiest camera-trap (Camera A), and then view the second camera-trap (Camera B), which covers the same spatial area. When watching footage on Camera B, ignore activity that was recorded by Camera A. Insert activity only seen on Camera B either as (i) completely new events, (ii) by turning single events from Camera A into a paired event based on the new information from Camera B, or (iii) by adding information seen on Camera B that was not on Camera A, such as the sex of the otter. For instance, Camera A may record an adult otter going into the structure with no paired exit. Camera B may record the exit, and also that it was a male. The single event recorded by Camera A can then be redefined as a paired event with the exit date and time and the sex of the otter.

6.5 Layer 2: otter information

6.5.1 Is it an otter?

An animal that is fully visible in the field of view is readily identifiable in the UK. The only other common species that could be mistaken for otter is an American mink *Neovison vison* (Figure 6.3), although there is potential for other mustelids, domestic pets and Eurasian beaver *Castor fiber*, and so species identification should consider all of these possibilities. Problems can arise if animals pass the camera-trap at very close quarters with only the fur visible, and/or if you cannot see the whole animal. The key difference in separating the species when you have a very limited view is the

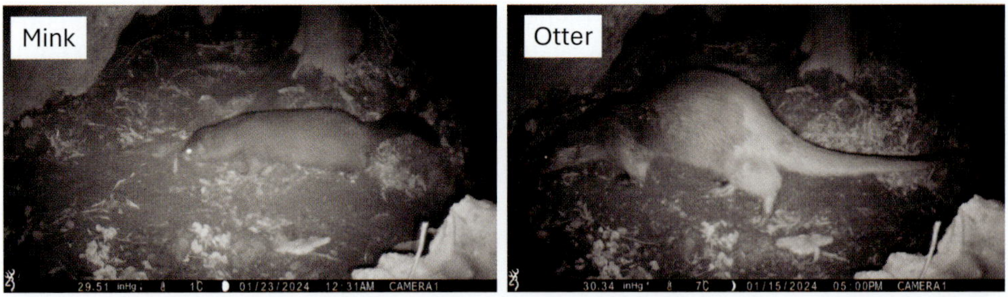

Figure 6.3 Comparison of American mink and otter from same site and camera-trap position.

Figure 6.4 Example of the colour and texture of fur in: (a) dry otter, (b) wet otter, (c) dry badger, (d) wet badger, (e) dry American mink and (f) wet American mink.

texture of the fur (Figure 6.4). When dry, an otter coat appears to be the smoothest, whereas the texture of the badger and mink is more obvious. The greatest difference, however, is when the animals are wet. An otter's coat is neatly covered in V-shaped points that are evenly distributed and look 'organised'. A closer view would show the very fine guard hairs forming a web-like covering. By contrast, both mink and badger fur clumps more unevenly and exposes the denser fur beneath, which is paler in badger (and can be in mink too, depending upon the colour variant of the mink).

Table 6.2 Distinguishing between otter and American mink on camera-trap footage. See Figure 6.3 for visual comparison.

Characteristic	Otter	Mink
Size and weight of adult (dependent on age, sex and condition)	Just over 1 m from nose to tail tip. Weight 7–9 (can be up to 11) kg	At 50–60 cm and 0.5–1.5 kg, approximately half the length and considerably less bulky than an adult otter
Colour	Evenly coloured with chocolate brown on top, paler underneath	Many variations from dark brown to pale grey (silver), can be patchy. Often has a distinct pale patch on the chin
Movement	The movements of mink are quicker and lighter in comparison with otter	
Tail	Tapered to pointed tip, tail at least 60% the length of the body	Furry, approximately half the length of the body, parallel-sided and blunt tip
Tail posture	Tail usually arcs downwards, see below, as it leaves the body, especially when moving, or can be straight when at rest 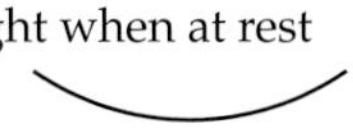	Tail tends to arc upwards as it leaves the body as below or is held straight 

As most captures are at night and therefore in black and white, tonal differences are important. A summary of some differences in characteristics between otter and mink is in Table 6.2. Comparison between otters and mink in the field is provided by Chanin (1993). Occasionally, a very poor view of a mammal can be captured that is impossible to say what species it is. In these cases, it's fine to say identification was not possible.

6.5.2 Determining sex

Being *sure* of the sex of the otter is only possible where you get a good look at the 'working parts' of the abdomen, either to observe genitalia or nipples or a stream of urine. This means that determining sex is far more likely with sexually mature adults. Large dog otters can sometimes be identified from their bulky shoulders, broad neck and head, but experience is needed before committing this to a datasheet. At our long-term study site we had a female that was very broad and could have been mistaken for a male. Likewise, we had a smallish but well-endowed male that with the wrong angle to the camera could have been mistaken for a female. Males tend to sniff a lot and are more likely to snort (hence it is useful to watch footage back with the sound on). Males are also more likely to be on their own (i.e. without being followed by cubs) and to loaf in the holt entrance and scent mark with ventral glands. This looks like they are rapidly rubbing the 'heels' of their front paws on the ground and rubbing their groin. These are all observations by the authors, however, and haven't been analysed.

Good camera-trap placement and settings can influence the ability to identify the sex of an otter. Setting camera-traps low down to show the urine stream can be definitive, with males' urine tending to emanate from further forward of the hind legs (Figure 6.5) while in females it emanates from further back and may stream backwards (Figure 6.6). When using a urine stream to sex otters, it is important that what you are looking at is really urination. Often, a small amount of liquid can shoot out backwards from the otter. This has been too rapid for our camera-trap footage to be able to discern whether this is scent marking by urine or spraint.

Figure 6.5 Male urination stream (shown with white arrow), which emanates from slightly further forward of the legs.

Figure 6.6 Female urination stream (shown with white arrow), which emanates from further back and may stream backwards.

Presence of a scrotum and/or prepuce are definitive features of an adult male. The prepuce can sometimes be seen in young males before they are sexually mature if the otter rolls onto its back. In this position, the scrotum can also be seen in sexually mature males. Both these features can often be seen in sexually mature males with a good, low side view (Figure 6.7). Of the two features, it is the scrotum that is seen most often with a low camera angle (Figure 6.8).

Figure 6.7 Side view of an adult male otter showing the scrotum between the legs and the prepuce, which is forward of this and is where to expect the origin of the urine stream in males.

Figure 6.8 Adult male with scrotum (shown with a white arrow) visible while walking. A low camera angle is useful to get such a view.

The female's four nipples on her lower abdomen become more obvious when she is lactating. These can be seen as pale patches in daylight (Figure 6.9), or as dark areas under infrared illumination (Figure 6.10).

It should be noted that still images are less likely to discern a urine stream and may also miss the split second when nipples or the scrotum are visible, and so video capture is better. It is important to concede defeat if there is not a good

Figure 6.9 Female laying on her back outside a resting place. The female is in lactation, and two nipples on the lower abdomen are prominent and visible as paler areas in daylight.

Figure 6.10 A female on her hind legs pulling at ferns to take into her holt as bedding. The female is in lactation and under infrared illumination the prominent nipple areas appear as dark patches (shown with white arrow), possibly due to wet fur from recent suckling.

Table 6.3 Categories that can be used when determining sex of otters.

Sex	Code	Confirmation
Female	F	Source of urine stream observed
Breeding Female	BF	Swollen or 'dropped' abdomen in conjunction with view of nipples
Male	M	One or more of the following: Presence of scrotum Presence of prepuce (if rolls onto back) Source of urine stream observed
Not Adult Male	NAM	An adult-sized otter where a clear view shows the lack of scrotum. Source of urine stream not observed. Potentially an immature otter or a female
Unknown	U	The sex can't be determined during a particular event if the view does not offer enough information

enough view to sex the otter, and often cubs are simply recorded as 'Unknown' sex. Some sites may mean that the camera-trap angle can't be set up to optimise the likelihood of sexing the animals. Suggested categories of sex for large otters are in Table 6.3.

There is less to see when trying to determine the sex of cubs. Observing the stream of urine can be used and if you get repeated observations of a family group you may see enough urination to confirm the sex of some of the cubs, if not all.

6.5.3 Determining age

Breeding females, and males with scrotums, are obviously adults. Likewise, very young cubs when they emerge from the natal holt can readily be aged at about eight to ten weeks from their appearance and behaviour. The cubs are usually chubby and fluffy on emergence and do not walk with a raised tail. This gives the impression that the tail is proportionally shorter in comparison to its body than an adult's tail/body ratio. While tail length may seem short in proportion to the body in cubs once they have emerged compared to adults, post-mortem measurements show only a very slight difference (Oliver 2023), which is too small to be useful when looking at cubs on footage. On footage of litters from our long-term study site, a pixel-measuring method was used to look at the body and tail length of growing otter cubs relative to each other and to their mother (Danes 2024). Three litters were monitored from when they first emerged from the natal holt, so we knew their approximate age. Based on these data we came up with some rules of thumb based on a cubs' relative size, appearance and vocalisation in order to put it in an age category (Table 6.4). By the time cubs are nine months old they are virtually indistinguishable from the mother, except for subtle differences in behaviour. With older litters, otter mothers appear to be more vigilant, scanning the area in front of the holt, and are less inclined to initiate play than the older cubs.

Table 6.4 The characteristics of broad age classes, which could be allocated a code in the events list such as 'Emergence', 'Small' and 'Medium'. If there is no size difference between the female and cubs, these would be classed as 'Large'. Note that call frequency can be extracted from camera-trap footage using free-to-use sound analysis software.

Characteristic	< 3 months Emergence	3–6 months Small	6–9 months Medium
Nose to tail length rule of thumb	Less than two-thirds of mother's length, just post-emergence	Two-thirds to four-fifths mother's length	More than four-fifths of mother's length
Development indicators	During this period cubs lack the looping otter gait, don't flex spine, coat is fluffy, mother may haul cubs about, cubs stay very close to mother		
Call pitch	Cub calls noticeably higher in pitch from mother		Any difference in pitch between juveniles at this age and adults not apparently noticeable to human ear

6.5.4 Identification of individuals

Occasionally otters will be seen with obvious differences, which allow confident identification when seen again. Unfortunately, these tend to be injuries. Examples include broken tails, scars, loss of eyes and loss of toes. Injuries such as a limp or tick marks can also be used but care is needed because such injuries or parasites may be temporary. Different females may be identified if they have differently aged litters, but this does not apply once both litters are approaching adult size. Equally, females may have different numbers of cubs in their litters but this becomes less clear as the otters get older, because they may not always rest as a family in the same structure. In addition, cub mortality can be high so a litter of three may become a litter of two, or even one. Misidentification of litters is therefore possible.

In Shetland, individual otters can be identified by unique patches of white on their throats, but these are rare on mainland UK. Mainland otters can have lip and/or chin marks that are paler than the surrounding fur (Figure 6.11). These show up particularly well in the IR light used by camera-traps, given a good facial image. The size and position of the lip marks vary: they can be narrow or deep, can vary in length (sometimes being little more than a spot) and can be different on either side of the mouth. It is not known how unique these markings are, so while they may be useful if a small number of otters are being monitored, a larger study may encounter similar lip/chin marks on different animals that may not be distinguishable, or there may be several animals that lack lip and/or chin marks.

Figure 6.11 Lip and chin markings (seen as pale areas) may enable some individuals to be identified. The otter here has a thin, continuous lip marking, which is relatively symmetrical.

Identification of individual otters is therefore extremely difficult, although with long-term study and a great deal of familiarity with the family group, the 'regulars' can often be identified due to their particular behavioural modus operandi, specific lip markings or a physical quirk, but caution is needed.

6.5.5 Adding information on otters to events list

With information on sex, age and possibly individuals, information on the otter(s) at a given structure starts to build. For example, in Figure 6.12, where information has been added in Layer 2, it can be seen that an adult male has visited

	A	B	C	D	E	F	G
1	**Explanatory**		**LAYER 1: Activity**			**Layer 2: Information about otters**	
2	OBS.	OBS.	OTTER IN	OTTER OUT	OTTER PASS	AGE	SEX
3	No.	TYPE					
4	1	Paired	23/10/2024 02:30	23/10/2024 02:30		U	U
5	2	Paired	23/10/2024 03:00	23/10/2024 03:02		A	M
6	3	Paired	24/10/2024 07:30	24/10/2024 17:30		A	F
7	4	Paired	24/10/2024 07:30	24/10/2024 17:30		MC	U
8	5	Paired			24/10/2024 17:00	A	NAM
9	6	Single	25/10/2024 07:00			A	U
10	7	Single		27/10/2024 13:50		A	U

Figure 6.12 Example Excel spreadsheet showing the events list populated with activity and information about otters observed. A = adult; F = female; M = male; MC = medium cub; U = unknown; NAM = not adult male (see Table 6.3).

the structure (observation 2), while a female and medium cub slept there later on (observations 3 & 4). Observations 6 and 7 could be either the male or female already observed, or other otters.

6.6 Layer 3: behaviours

6.6.1 Bedding collection

While bedding collection sounds like an obvious behaviour to observe, it can be missed quite easily because the otter usually moves much faster than normal when collecting bedding. You may only get the tip of a tail on your camera-trap footage or even miss the otter completely. From our experience, one or two videos will capture the otter rushing past with a mouthful of bedding. During bedding collection the otter can zoom in and out and avoid detection because camera-traps are more likely to miss fast moving animals. It is useful to turn to the Trash folder for any false triggers around the same time as you have a record of the otter with a mouthful of bedding. A second examination of the videos at this time may reveal rummaging noises from out of view, or a disappearing tail end may be spied that wasn't noticed before. These missed triggers will enable you to record a more accurate duration of bedding collection.

Bedding collection could be recorded on your events list as a separate 'Bedding collected' column with simple YES/NO categories. However it is better to record the duration of bedding collection in minutes either by working it out yourself from the start and end time of the bedding collection activity or by logging the start and end times of bedding collection in separate columns and calculating the duration in Excel. The duration of bedding collection can be evidence of a natal holt (see section 6.7.1).

6.6.2 Latrine behaviour

Both urine and faeces are deposited in latrines and often this behaviour can be observed on camera-trap footage. It represents a great opportunity to sex the otter as it urinates. The latrine can be used prior to, during, or after a rest, and the otter makes a meaningful deposit, unlike spraints, which are swiftly deposited and very modest by comparison. During a long rest, it is usual to see the otter wake up and use the latrine if visible, often several times. During our observations in Scotland, otters generally did not spraint on their latrines when visiting the site, and spraint sites were usually separate. To fully understand the difference between spraint sites and latrines as field signs (see section 2.2).

Latrine behaviour could be recorded on your events list as whether or not it occurred (YES/NO) or the number of latrine visits could be counted.

6.6.3 Stationary behaviours

Often at a resting place, the otter or otter family will spend time at the entrance to the structure if there is a dry area immediately outside. Grooming, loafing, pulling at vegetation, play fighting, scent marking and sleeping are all possible. These can be captured by a simple measure of number of minutes they were seen for. As this is not essential to identifying resting places or breeding places it is not necessary to record, but it may be of interest.

6.6.4 Other behaviours

You may wish to record other behaviours you think are important, such as food items being taken into the resting place (very rare), sprainting, eating (very rare) and vocalisation. These may be of interest to an individual study but are not central to determining whether the site is a resting or breeding place.

6.6.5 Adding behaviours to the events list

Recording behaviour highlights some registrations as potential rests. For example, observation 7 in Figure 6.13, a single event, becomes of more interest when we note that the otter used the latrine on the way out of the structure, so is likely to have rested.

	A	B	C	D	E	F	G	H	I
1	Explanatory		LAYER 1: activity			Layer 2: information about otters		Layer 3: behaviours	
2 3	OBS No.	OBS. TYPE	OTTER IN	OTTER OUT	OTTER PASS	AGE	SEX	BEDDING (min)	LATRINE (count)
4	1	Paired	23/10/2024 02:30	23/10/2024 02:30		U	U	0	0
5	2	Paired	23/10/2024 03:00	23/10/2024 03:02		A	M	0	0
6	3	Paired	24/10/2024 07:30	24/10/2024 17:30		A	F	3	1
7	4	Paired	24/10/2024 07:30	24/10/2024 17:30		MC	U	0	1
8	5	Paired			24/10/2024 17:00	A	NAM	0	0
9	6	Single	25/10/2024 07:00			A	U	0	0
10	7	Single		27/10/2024 13:50		A	U	0	1

Figure 6.13 Example Excel spreadsheet showing the events list populated with activity, information about otters observed and their behaviours.

6.7 Types of activity at resting places, characteristics of natal and maternal resting places

The following section provides context or 'expectations' of what would be recorded at different types of resting place. It is important to distinguish natal and maternal resting places from other resting places because these breeding places are likely to be a rarer resource in the female's home range. It is important to remember that a resting place can be multi-functional. For instance, a natal resting place can later be used for resting by the female and older cubs, or even by other otters. Of principal importance is to understand that a site can only be proven as a natal resting place during the eight to ten week period before the cubs emerge. If a camera-trap is set up and young cubs come out on the first night, it can't be assumed that the cubs had been born there.

6.7.1 Natal resting places

The following information is based on observations of litters from a natal holt in a drift mine in Scotland, which was repeatedly used. Little information, aside from our observations below, is available in the wider literature on what to expect at a natal holt, other than a radio tracking survey of a female otter that gave birth during the study period (Durbin 1996). This study found that just before giving birth the female moved to a natal holt away from her normal foraging area along a small watercourse. Her activity decreased after birth for approximately three weeks, then increased, leaving the cubs for extended periods, of up to 8 hours and 20 minutes, while she was foraging.

Our study structure is perhaps unique in that many otters can rest in it simultaneously, but we believe it has a number of separate chambers and so those otters may be able to occupy different parts of the structure. Although this study holt may be unique, it challenges many of the preconceptions about natal holts, and the key message, therefore, is to keep an open mind because there is still a lot about natal holts to be learned.

Female – appearance and activity

It is variable when the female moves in to occupy and prepare the natal site. Generally, this may be just before the birth, although we had one female that occupied a natal holt several weeks before giving birth. Once the cubs are born the female lives at the holt and makes daily trips away to forage. If the camera position is good, the swollen belly and teats of the pregnant female can be seen. After the cubs are born the abdomen can look 'lumpy' while she is nursing. After feeding the cubs, the area around the teats can appear dark and is presumably still damp after the cubs have been suckling. It is thought there are between eight and ten weeks between birth and the emergence of the cubs.

Behaviour – bedding collection

The time of birth is usually marked by increased bedding collection. This has been noted in captive otters (Wayre 1979) and we have noted up to 35 minutes of bedding collection, which contrasts with the much shorter periods, usually less than ten minutes, that are not associated with birth or the presence of young cubs. Long periods (> 10 minutes) of bedding collection by a female, especially on several consecutive nights is likely to indicate the presence of young cubs or impending birth. Males also collect bedding material, but these sessions usually comprise just one or two return journeys out of the holt to collect vegetation. Often this is after a rest, even if the otter does not return to the holt for the following diurnal rest.

Food and eating

We have not recorded food items being taken into the holt until the cubs are just about to emerge. At this point, the female may take an item of food into the holt but it is a rare occurrence and emergence is imminent. If a female is taking food into cubs she may growl or mew, presumably to inform the cubs of impending food. A single otter (without cubs) may also take the odd item of food into a holt. This is also very rare and we did not observe vocalisation in these very few cases.

Sharing of the natal holt

In the natal holt we studied, other otters often rested in there during the natal period. In one natal period, the male cub from the previous year's litter regularly rested in the holt, as did a large adult male. In every natal period we have monitored, a male would visit the holt, scent mark, and either rest in the structure or simply visit without resting. One year, two females used the holt at the same time for giving birth and raising their cubs for the first eight to ten weeks. One litter appeared 12 days before the other. While co-habitation of natal holts may be rare, the presence of other otters, even adult males, should not be taken to indicate that the structure isn't a natal holt.

6.7.2 Maternal resting places

Maternal resting places are where the young cubs are reared until they are six months old (see section 3.2.2). Once otter cubs have emerged from the natal holt they are usually moved from resting place to resting place and it is useful to distinguish these as maternal resting places. The natal holt may also be used as a maternal holt to which the female and cubs occasionally return for a few days at a time, but this is likely to vary depending upon where the structure is in relation to good foraging areas. After emergence, the cubs still rely heavily on the security of the maternal resting places while the female forages. The cubs continue to be fed by the mother until they are at least 14 weeks old (Wayre 1979). We have observed a female with suckle marks and with a cub ten weeks after emerging from the natal holt. The cub was estimated to be 20 weeks old (i.e. between 5 and 6 months).

During these times, the cubs can be observed in the holt entrance play fighting and softly chittering to each other, pulling at vegetation or simply investigating their surroundings. There are also likely to be periods when the female loafs at the holt entrance with the young cubs. These periods of activity can seem very extended and can trigger video after video on a camera-trap, but when their duration is examined they often last just a few minutes. The cubs increasingly accompany the mother away from the holt as they get older. Bedding collection may be observed at maternal holts and, again, food items are very rare. Other otters visiting or resting in the structure should not be taken as an indication that the structure isn't a maternal holt.

6.7.3 Resting places in general

We analysed camera-trap data collected across a large number of potential resting places in the River Tweed catchment (Findlay et al. 2023) and from the long-term study of a holt in central Scotland (Findlay et al. 2017). We found that otters would frequently 'visit' all these sites, spending a very short period of time (often less than a minute) at the site, when they would sniff and perhaps spraint, possibly not even going into any structure if present. However, at a proportion of these sites the otter would remain for much longer (as indicated by paired events). In effect, they were resting, and at one structure the otter could be seen sleeping within.

When we looked at the distribution of time spent at the site or in the structures from the 26 River Tweed study sites as shown in Figure 6.14, most of the registrations

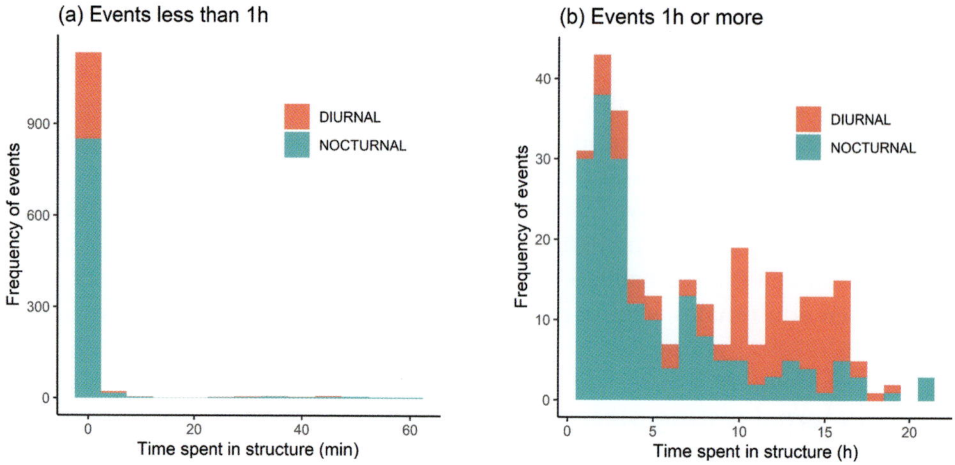

Figure 6.14 Distribution of time otters stayed inside structures that were identified as potential resting places, based on c. 1,400 paired events: (a) events less than one hour, and (b) events one hour or more. These data helped define a useful rule of thumb to distinguish between resting places and structures not used for resting. From Findlay et al. (2023) (CC BY 3.0).

were either short visits of a minute or less (Figure 6.14a) or obvious rests of several hours (Figure 6.14b), without much in between these peaks in the data. We therefore adopted the threshold of 15 minutes or longer to define a resting event. This was based on the data and also on common sense: if an otter is in a holt for longer than 15 minutes, it is likely to be inactive or resting.

The Tweed data also showed that some rests were diurnal and others nocturnal. Both types of rest were recorded at all resting places. In general, nocturnal rests tend to be shorter than diurnal rests. The median rest duration of a diurnal rest was 12.3 hours, compared to the much shorter median duration of a nocturnal rest, which was 2.7 hours, but most of these rests were during the winter and spring. Summer/autumn rests may fall within a different range.

Now that a time limit of a 15 minutes in the holt can be considered a good threshold with which to pull out rest events from the data, the structure can be classified as a resting place if one or more rests are observed. The number of rest days can also be reported as a proportion of the camera-trap days. It should be stressed that an adequate number of camera-trap days is needed before a site can be discounted with a reasonable degree of confidence as a resting place (see section 5.5.2).

6.8 Other scenarios when using camera-traps for otters

6.8.1 Terrestrial routes

Camera-trapping of paths leading to features such as ponds, ditches or that cut across meander loops needs to be very clearly thought through. If there is some doubt as to whether it is an otter path or a badger path, and the path will be affected by a development, camera-trapping may be necessary to confirm use by otter. The difficulty lies in how to interpret the data. How do you decide when a path is important enough to mitigate any effects? One otter pass per week? One otter pass per night?

In terms of sampling and camera-trapping effort, it would be wrong to assume that the path has the same importance all year round and so sampling each season would therefore be needed. Putting out a camera-trap for a couple of weeks is not likely to be very informative. If the path leads to an important resting or breeding place but you can't pin down the exact location, more frequent activity at sunrise and sunset may provide evidence of diurnal resting nearby. If an otter path is clearly an otter path based on field signs, it will have been formed from regular otter footfall. Camera-trapping may not offer a great deal more information, and it will be difficult to justify the need to clients. In short, think twice before camera-trapping a terrestrial route and be clear in your mind what the data could look like, how you would interpret and apply it to your project, and what your sampling pattern would be. Ask whether the resulting information will be more than what you could achieve using simple observation – that it was an otter path and was likely to be regularly used.

6.8.2 Non-resting places

Monitoring spraint sites can offer information on the otters in the vicinity, but inferring anything about the number of otters is not possible, because many animals cannot be recognised as individuals in the same way that pine martens can be recognised from pelage patterns. Equally, it is not possible to state anything conclusive about the level of activity or compare otter activity at different locations, because the camera-trap samples such a small area.

It can be beneficial to know the breeding status of females and whether they have cubs and, if so, the age of the cubs. In this case, it is best to monitor several spraint sites simultaneously and set the camera-traps very low down to get a good look at the otters' working parts, to be better able to assess sex and breeding status.

6.8.3 Monitoring otter 'activity' during construction

First define what you mean by 'otter activity', and then how you could measure it with a reasonable level of certainty. As discussed in section 2.5, spraint surveys are not adequate to assess otter activity. As camera-traps only monitor a few square meters of ground, having no otter registrations on a couple of camera-traps would not demonstrate the absence of otters. The only way that reasonable conclusions can be drawn about the effects of construction activities would be if there had been a good baseline camera-trap survey of a particular feature such as a resting place or path. Setting up camera-traps in the same location as in the baseline survey would allow a comparison of registrations per week or similar, but care would be needed to account for any variances in season or other differences.

6.8.4 Monitoring artificial resting places

The interpretation of camera-trap data from artificial resting places should be the same as for natural resting places. The events list and paired registrations showing that an otter has been inside the structure for longer than 15 minutes would be evidence that an otter has rested there. Simple presence of otter at an artificial holt is not evidence of resting.

6.8.5 Ledges and culverts

Provision of dry tunnels (dry culverts) under roads to provide a safe alternative route for commuting otters when the river is in spate is a common method to reduce road mortality. Fitting ledges under existing bridges retrospectively has also been used for the same reason. Proof of use of a dry culvert is straightforward: the otter must be sequentially registered by camera-traps at both ends of the tunnel. During spate conditions, that would be a good indication that the culvert is serving its purpose. Proof that a ledge functions as successful mitigation is more difficult because an otter could choose to swim in the channel and pass undetected in the water, may use the ledge but due to its wet fur it may not trigger the camera-trap or the otter may simply head back from where it came rather than face the ledge or the (fenced-off) road. Standalone camera-trapping of otter passes along a ledge during spate conditions may not be the best method of proving that the mitigation works. However, coupling this with monitoring otter mortality at the road crossing could theoretically demonstrate that otters are not necessarily impeded by the tunnel during spates and that the road tunnel is not a cause of mortality.

6.9 Quality control

It is good practice to look over your data to check that you are happy with it or at least understand or can account for any idiosyncrasies. Some simple checks can tell you much about the quality of your camera-trap set-up and whether you need to be cautious about interpreting the data. This is only possible at resting places where you are as confident as you can be that there is only one entrance or, if more than one entrance, that you are camera-trapping all entry/exits.

6.9.1 Ratio of single to paired events as an indicator of data quality

An excessive proportion of single events (section 6.4.1) may be due to suboptimal camera-trap set-up in relation to detection probability. This may be due to inexperience or because the optimal set-up isn't possible. Nonetheless, the proportion of single events to paired events is important and may suggest that there is another entrance to the structure that you are unaware of (such as an underwater entrance) or an entrance that you are not monitoring for another reason, such as (i) it can't be accessed safely or (ii) there are no opportunities for a camera-trap to be deployed because the entrance is in a pile of rocks that won't accommodate a wooden stake and there are no natural fixing points to exploit. It is unrealistic to expect 100% paired events because camera-traps are not infallible; from our experience, a good set-up may show that 80% or more of all events are paired.

If you have an excess of single events, first ask whether there is a pattern, for example are they are biased to the animal leaving or entering the holt? If the holt entrance is close to water, you may have missed otters leaving the water and entering the holt because the otter has wet fur and is less likely to trigger the camera-trap, especially if the otter tends to go straight into the holt without pausing to shake or spraint. The higher rate of capture of otters leaving the holt compared to entering may thus be due to the animal being drier and warmer, or perhaps pausing between leaving the holt and entering the water. If there is a bias towards recording otters entering the holt, it may be that the otters pause or loiter more before they go in but on the way out swiftly enter the water and do not give the camera-trap enough stimulus to trigger, or they do trigger the camera-trap but are out of the field of view before it starts recording.

If you identify such a bias it would be useful to examine the times of the single events in relation to sunrise (for holt entries) or sunset (for holt exits). Diurnal rests are characterised by otters exiting the holt at about sunset or shortly afterwards, and otters entering the holt at about sunrise or shortly before. We must caveat this slightly in that most of our holt-use data is from winter, when the nights are longer. There may not be such a strict pattern in summer, because the short nights may not give the otter enough foraging time and so activity may spill into the daylight hours, especially in less resource-rich habitats.

When an otter exits a holt at about sunset it can be useful to note whether its fur is completely dry and fluffy. This is likely to indicate that it has been in the holt long enough to properly dry off. Some caution is needed if there is a possibility that there is water within the holt structure. It is also worth noting the otter's behaviour. After a rest an otter will frequently lie in the entrance of the holt and act 'sleepy'. It may yawn or shut its eyes for a mini-nap of a couple of minutes. Thus, a pattern of single events that aligns with sunrise/sunset can be enough evidence to state that the structure is likely to be used for diurnal resting, supported by dry fluffy fur on exit and/or sleepy behaviour.

Nocturnal rests with incomplete data are much harder to pin down. Such rests tend to be shorter in duration, but again dry fur and a sleepy countenance on exiting the holt should be a flag for a nocturnal sleep to be considered.

6.9.2 Funky exposure

Despite the most careful set-up, it is possible for little bits of vegetation to end up sitting in front of and close to the camera-trap. These can be blown bits of debris, a fallen leaf or vegetation dislodged by a badger, or even by an otter when collecting bedding. This can cause the emitted illumination from the camera-trap to reflect back at the camera and automatically cause the camera to adjust to this reflected light resulting in a bright, white frond of vegetation within a black image on the thumbnail (Figure 6.15). However, there will usually be one or two frames at the beginning of a video, before the camera reacts, when the otter and its direction of movement might be seen. Daytime footage is barely affected other than partial blocking of the field of view by the vegetation.

6.9.3 Time in holt

The duration of a diurnal rest tends to vary but is generally between 6 and 16 hours (see Figure 6.14). Rarely, longer rests occur where an otter enters a holt in the middle of the night and doesn't emerge until after dusk on the following night. At our long-term study site we noted that the longest rests were by males and tended to be the first in a

Figure 6.15 The effect of vegetation close to the camera-trap (using a Browning Spec Ops Elite). The upper image shows the first frame of the video prior to any auto adjustment to the vegetation, and the lower image shows the sixth frame where the exposure has been adjusted and the otter is not visible. This frame forms the thumbnail.

series of two or three consecutive diurnal rests. Our interpretation was that the male had travelled some distance to get to the holt and that his first rest allowed a recovery.

Any paired registrations where an adult otter is in the holt for longer than 16 hours should be carefully validated. False triggers between the entry and exit should be rechecked to see if the two otter registrations are actually related to the same event. It could be that registrations that have been paired and that indicate that the otter has been inside for longer than 16 hours, are due to incomplete data rather than a single long rest, and that an otter exiting the structure and re-entering was somehow missed. If necessary, the pairing should be split into two single events to avoid over-interpretation of the structure.

6.9.4 Position of the otter in the field of view in the first video frame

It is worth noting the position of the otter in the first frame of the video, which represents the closest record of where it could have triggered the camera-trap. Sometimes the otter can 'appear' in the centre of the footage without any capture of where it came from. It could have come out of the holt or run in from out of the field of view. In these cases, first assess whether it is feasible for the otter to have triggered the camera-trap at that point. For instance, suppose a camera-trap faces an entrance to a structure and the first video frame shows an otter tucked into the back of the structure with just part of its face showing and facing outwards. Is it reasonable to think that this otter triggered the camera-trap in that position, or could it have gone into the structure and turned around before the camera-trap triggered, which would make it appear to be exiting (rather than entering and exiting)? A second possible explanation is that there are two otters: the first one exited and triggered the PIR but had already scooted past by the time the camera switched on and you have the second otter in frame. It takes a bit of experience to pick up on these instances, but if this were the only registration of otter, it is an important consideration.

6.9.5 Is the otter present in the first video frame?

If the otter appears to come out of a structure part way through a video (Figure 6.16) something else has triggered the camera-trap. It would be unlikely for waving

Figure 6.16 This first view of the otter (eyeshine visible in lower centre of frame) is after 10 seconds of video footage. The trigger could have been an otter coming out prior to this one and dashing past, or it could have been a late trigger when the otter on the screen entered the holt and so the entry was not recorded.

vegetation or a mouse to trigger the camera-trap at such an exact moment (although never say never!), but it is more likely that the cause of the trigger has not been recorded. This could be another otter dashing past or the camera-trap may have triggered as the otter entered the holt, but it was a late trigger and so this was not recorded.

6.10 Reporting, summarising and interpreting your data

6.10.1 What do you need to find out?

Now that the events list is complete the next step is to interpret and summarise what you have found. The most common aims for an otter survey are:

- Broad summary: how many days were otters recorded?
- Is the structure used as a resting place?
- Is the structure being used as a natal resting place?
- Is the structure being used as a maternal resting place?
- Is the structure being rested in by more than one otter?

6.10.2 Reporting the method

It goes without saying that this is a crucial part of a camera-trapping report. Accurate reporting of the set-up is particularly important for an otter camera-trapping study as each set-up is likely to be bespoke to the site, situation and proposed development. The checklist in section 5.6.5 is a good summary to include in the report appendix.

6.10.3 Simple tools in Excel

Adding formulae in Excel may seem like a real faff to some. However, with busy sites where you want to see if there are any rests and therefore classify according to whether it is a resting place or not, data that is well set out in a spreadsheet helps you to readily pick out the occasions where rests have occurred and quantify them too. Adding a few columns and calculations in Excel will really start to show patterns that facilitate summary statistics suitable for reporting.

Remember to check that the date and time columns are correctly formatted before you start. The duration of time spent in the holt can be calculated by subtracting the time in from the time out for each paired observation. Errors in entering the date or time may show up as very long rests or as negative numbers, so check these. Secondly, you can add an identifier to all observations where an otter was in the holt for longer than 15 minutes or however long you wish to define a rest (Figure 6.17). To do this add two new columns, 'Time in holt (min)', and 'Rest'. The formulae using time and dates are not easy to grasp and are given in Figure 6.18. Don't forget that behaviour can be used as evidence of a rest too, so a single event associated with use of a latrine can also be flagged as a rest. Once these columns have been calculated you can use the 'Filter' tool in Excel to look at all rests, or you may be interested only in rests by females and cubs, or by males. If Excel formalae are not your cup of

	A	B	C	D	E	F	G	H	I	J	K
1	Explanatory		LAYER 1: activity			Layer 2: information about otters		Layer 3: behaviours		Use formula to work out	
2	OBS	OBS TYPE	OTTER IN	OTTER OUT	OTTER PASS	AGE	SEX	BEDDING	LATRINE	Time in holt (min)	Rest (YES or NO)
3	No.							(min)	(count)		
4	1	Paired	23/10/2024 02:30	23/10/2024 02:30		U	U	0	0	0	NO
5	2	Paired	23/10/2024 03:00	23/10/2024 03:02		A	M	0	0	2	NO
6	3	Paired	24/10/2024 07:30	24/10/2024 17:30		A	F	3	1	600	YES
7	4	Paired	24/10/2024 07:30	24/10/2024 17:30		MC	U	0	1	600	YES
8	5	Paired			24/10/2024 17:00	A	NAM	0	0		NO
9	6	Single	25/10/2024 07:00			A	U	0	0		NO
10	7	Single		27/10/2024 13:50		A	U	0	1		YES

Figure 6.17 Using spreadsheet formulae allows calculations such as the time the otter spent in the holt (column J). It is then possible to flag rest events (column K).

	A	B	C	D	E	F	G	H	I	J	K
1	Explanatory		LAYER 1: activity			Layer 2: information about otters		Layer 3: behaviours		Use formula to work out	
2	OBS	OBS TYPE	OTTER IN	OTTER OUT	OTTER PASS	AGE	SEX	BEDDING	LATRINE	Time in holt (min)	Rest (YES or NO)
3	No.							(min)	(count)		
4	1	Paired	23/10/2024 02:30	23/10/2024 02:30		U	U	0	0	=IF(COUNTBLANK(C4:D4)=0,(D4-C4)*1440,"")	=IF(OR(NOT(OR(J4="",J4<15)),SUM(H4:I4)>0),"YES","NO")
5	2	Paired	23/10/2024 03:00	23/10/2024 03:02		A	M	0	0	=IF(COUNTBLANK(C5:D5)=0,(D5-C5)*1440,"")	=IF(OR(NOT(OR(J5="",J5<15)),SUM(H5:I5)>0),"YES","NO")
6	3	Paired	24/10/2024 07:30	24/10/2024 17:30		A	F	3	1	=IF(COUNTBLANK(C6:D6)=0,(D6-C6)*1440,"")	=IF(OR(NOT(OR(J6="",J6<15)),SUM(H6:I6)>0),"YES","NO")
7	4	Paired	24/10/2024 07:30	24/10/2024 17:30		MC	U	0	1	=IF(COUNTBLANK(C7:D7)=0,(D7-C7)*1440,"")	=IF(OR(NOT(OR(J7="",J7<15)),SUM(H7:I7)>0),"YES","NO")
8	5	Paired			24/10/2024 17:00	A	NAM	0	0	=IF(COUNTBLANK(C8:D8)=0,(D8-C8)*1440,"")	=IF(OR(NOT(OR(J8="",J8<15)),SUM(H8:I8)>0),"YES","NO")
9	6	Single	25/10/2024 07:00			A	U	0	0	=IF(COUNTBLANK(C9:D9)=0,(D9-C9)*1440,"")	=IF(OR(NOT(OR(J9="",J9<15)),SUM(H9:I9)>0),"YES","NO")
10	7	Single		27/10/2024 13:50		A	U	0	1	=IF(COUNTBLANK(C10:D10)=0,(D10-C10)*1440,"")	=IF(OR(NOT(OR(J10="",J10<15)),SUM(H10:I10)>0),"YES","NO")

Figure 6.18 Formulae to calculate the time the otter spent in the holt (column J).

tea, you can do this step manually and type YES in the rest column if either (i) time in holt is more than 14 minutes or (ii) you have detected any bedding collection or latrine use. From here, it is an easy step to summarise your data.

6.10.4 Producing a summary of otter activity for a report

This should start with your metadata – the number of potential camera-trap days through the study and the number of actual camera-trap days achieved – that is, the number of days during which the camera-traps were functioning.

Otter presence can be expressed as the number of days that an otter was recorded, either as a figure or a percentage of the actual camera-trap days. Be wary, however, that if there were several visits to the structure on the same night, these would only count as a single day when otters were recorded. You are unlikely to know for each of these multiple-visit nights whether different otters were visiting or the same otter kept returning. It's best to keep it simple – to the proportion of days when one or more otters were observed.

The *number of rest days* can be expressed as the number of days that an otter was recorded in the structure for 15 minutes or more (or however you like to define a rest), either as a figure or a percentage of the actual camera-trap days. It may be that the structure is rested in by different otters at different times, such as an adult male and a female with two small cubs, and it would be useful to split the rests according to the otter(s) resting.

Comments on behaviour are useful as observations, but if significant, such as frequent bedding collection, quantifying the time spent would be informative.

Comment on the individuals, such as an adult male was recorded, a female with two young cubs and a female with one large cub.

Once you have summarised your data in this way (Table 6.5) you can start to draw conclusions, but it is also useful to cast your eye over the data to see if you think it is reliable.

Table 6.5 Example of how results can be summarised from a camera-trap survey at a potential resting place.

Summary statistic	**Days**	**% of actual camera-trap days (n = 98)**	**Notes**
Potential camera-trap days	130	n/a	
Actual camera-trap days	98	n/a	Battery depletion, camera-trap functioned for 75% of days
Otter presence	80	82	
Otter diurnal rests – adult male	4	4	
Otter diurnal rests – female and two cubs	9	9	Short bouts of bedding collection on two rest days
Otter nocturnal rests – female and two cubs	1	1	Female arrived with cubs and left them for three hours to forage alone
Otter diurnal rests – unknown	1	1	
Otter rests – total	14	14	

Interpreting our findings from Table 6.5, we can confirm that the site is a resting place and a maternal resting place because a female with young cubs rested there (it should be noted that cubs were left unattended at the site). Natal activity wasn't observed during the camera-trap monitoring.

6.11 Key points

- Understanding how to collate, interpret and present your camera-trap data is crucial for any survey work. Excel spreadsheets are a useful tool for data to be recorded in, summarised and presented.
- Creating a calendar or timeline of activity, called an Events List, is recommended to determine activity and resting behaviour. Evidence of an otter resting is when the date and time of an otter entering and leaving a structure or place is recorded and the duration spent at the structure or place is at least 15 minutes. Use of the latrine is also evidence of a rest, even for a single event.
- Recording the age, sex and behaviour of otters at resting places can be useful to ascertain resting and breeding.
- Resting and breeding places are also characterised by bedding collection and latrine behaviour, although the latter may not be visible on camera-traps if the latrine is hidden inside the structure.
- Activity patterns at natal resting places are characterised by daily outings of a female for at least eight weeks.
- An intensive period of bedding collection (i.e. 10 minutes continuous) can be a sign that a female has, or is about to, give birth.
- Use of a place for resting by a female and cubs of up to six months old characterises a maternal resting place, and there are likely to be many maternal resting places in the female's core range.
- Camera-traps can be used to monitor otters on terrestrial routes and at artificial resting places, on ledges and in culverts, but the survey must be carefully designed and have clear aims and rationale.
- It is good practice to review the quality of any camera-trap data to understand any limitations of the data.

Glossary

Anal jelly: a jelly or marmalade-like substance deposited at spraint sites that is thought to be part of the mucous lining of the gut, which protects the gut from sharp fish bones.

Black IR: an informal term used to mean infrared (IR) illumination from camera-traps that is not visible to humans (and probably other mammal species). Black IR LEDs typically emit at a peak 940 nm wavelength.

Breeding season: the phenomenon whereby breeding of an animal tends to occur entirely or primarily during a particular period or season of the annual cycle.

Breeding site: a term taken from the Habitats Regulations meaning the habitat that supports the mating, birth and rearing of young.

Breeding place: a resting place that is used for birth of cubs (natal resting place) and/or rearing of small cubs until they are six months old (maternal resting place).

Camera-trap: a standalone device designed to take images or short video clips when triggered by the motion of an animal in front of the camera. While models vary, common features include passive infrared (PIR) triggering and infrared illumination to allow recording at night.

Core range: for otters, this is the area of the female's home range in which she gives birth and rears her young cubs.

Couch: often used to describe resting places that are above ground, such as in dense seasonal vegetation, gorse scrub or on riverbank ledges.

Detection probability: in the context of camera-trapping, the probability that a camera-trap captures video or an image of a target animal, assuming that it is present within the detection zone. Broadly speaking, this has two components: trigger probability and registration probability.

Detection zone: the area in front of a camera where an animal can potentially trigger the camera-trap and be detected by it in the form of images or video clips.

Disturbance: defined by the European Commission (2021) in guidance relating to the Habitats Directive as an act that impacts 'the chances of survival, the breeding success or the reproductive ability of a protected species, or that leads to a reduction in the occupied area or to a relocation or displacement of the species'.

Duty time: where a camera-trap is set up to record during specific times of day, typically as a means of reducing the number of non-target triggers.

Emplacement: in the context of camera-traps, any additional item that you place in the environment, such as a stake, a beam or a bracket that gives you the optimal location for your camera-trap, as opposed to strapping it to the closest available tree or other natural object.

Events list: a calendar of relevant events observed via one or more camera-traps during a camera-trapping survey at a site such as a holt. It shifts the unit of recording from individual camera-trap video clips or images to meaningful biological events that may have been recorded across multiple video clips or multiple camera-traps at a site.

False trigger: any camera-trap footage or images triggered by non-target stimuli, including vegetation movement, rainfall or other species.

Field of view: in the context of camera-traps, the area that is potentially visible to the camera-trap. The field of view may be reduced by obstacles such as vegetation, and at night the IR illumination of a camera-trap may not fill the whole daytime field of view.

Field signs: evidence left by otters in the field that may help us determine their presence or activity, such a spraints, latrines, sign heaps, rolling pitches, footprints, paths, slides, food remains or hair.

Haul-out: a term that has been adopted by some practitioners and is likely to have been borrowed from sea otter or seal terminology. Accepted definitions of this term in the context of otters can't be found, although it was assumed that these would be areas where an otter can lie by the side of the water while they eat and/or groom. This term is not recommended in survey reports due to its ambiguous definition.

Holt: an otter's enclosed, underground resting place, such as in burrow systems in riverbanks or peatlands, in hollow trees or in any enclosed structure.

Hover: a term that has been adopted by some practitioners more recently, but definitions vary and it is a non-standard term. In times gone by, hunters used this term to describe any place frequented by otter. This term is not recommended in survey reports due to its ambiguous definition.

Infrared (IR): the part of the electromagnetic spectrum with wavelengths longer than light visible to humans, typically from 780 nm to 1 mm wavelengths. Thus, it falls just adjacent to red light.

Infrared LED: a light-emitting diode (LED) designed to emit fully or mostly within the infrared part of the electromagnetic spectrum.

Invasive survey: in the context of otter resting places, this means using a torch, flash photography or camera equipment to survey inside a holt structure to look for evidence of resting such as bedding or latrines.

Inverse square law: in the context of radiation, a scientific law that states that the intensity of radiation is inversely proportional to the square of the distance of its source. In the context of camera-traps, which typically rely for their triggers on infrared radiation of animals, this means that distance from camera-trap plays a very important role in trigger probability.

Latrine: a feature comprising numerous droppings that are digestive eliminations as opposed to the token amount of faecal matter in an average spraint. The droppings accumulate to form a significant pile, which attains height and/or has horizontal spread, often covering an area as large as, and sometimes larger than, a dinner plate. Latrines are associated with habitually used resting places.

Licence: in the context of otter surveying, this is a document issued on behalf of the government that allows activities that are otherwise prohibited.

Litter: the cubs all born at the same time and raised together by the female.

'Low-glow' LEDs: LEDs that emit red infrared light.

Maintenance visit: a visit to a camera-trapped site between set-up and take-down to change SD cards and batteries and check that the camera-traps are functioning and still in place.

Maternal holt/maternal couch/maternal resting place: place or structure used by a reproductive female subsequently to a natal den, where the weaning of the kits/pups generally takes place until the cubs are approximately six months old. A female will use a number of maternal dens.

Mill lade: human-made channel that carries water to a mill and returns it to the river, also called a mill leat, mill run or mill race.

Natal holt/natal couch/natal resting place: place or structure where cubs are born. A female can only use one per breeding cycle.

'No-glow' LEDs: LEDs that emit black infrared light.

Potential resting place: a site that has been identified as having potential to be an otter resting place but is not yet confirmed as a resting place from field or camera-trapping evidence.

Prepuce: the covering of skin that protects the penis.

Rest: although not well defined in legislation, we take a rest to mean an otter spending more than just a cursory period at a site. For holts, we found that 15 minutes was a good threshold to distinguish instances where otters briefly visited holts from instances where they spent considerably more time in the structure.

Resting place: any place or structure used by stationary otters between moving bouts, for resting and/or sleeping (prerequisite).

Red infrared: an informal term used to mean infrared (IR) illumination from camera-traps that is visible to humans (and probably other mammal species) in the form of a red glow. Red IR LEDs typically emit at a peak 850 nm wavelength, which is close to the threshold with visible red light (780 nm).

Registration probability: in the context of camera-trapping, the probability that a camera records usable video clips or images of an animal, assuming it triggered the camera-trap. It is useful to separate this component of detection probability because it is influenced by different factors to trigger probability.

Passive infrared (PIR) detection: the standard motion detection mechanism of most camera-traps that detects differences in infrared (i.e. heat) radiation from surfaces, with the intention that an animal will be warmer (or sometimes cooler) than the surrounding background environment and will thus trigger the camera-trap.

Rolling pitch: an area in which otters roll as a means of drying and cleaning their coat and potentially scent marking. The rolling action often leaves an area of clearly flattened and/or abraded vegetation. They are often marked with spraint, and in coastal areas are usually associated with a freshwater source such as a small pool.

Sign heap: a structure formed by otters by scraping up small piles of vegetation, silt or sand on which to spraint or urinate, thus having a scent marking function.

Slide: erosion of a bank or slope where the otter enters or gets out of the water regularly. Other species also get in and out of the water, especially dogs, and so other corroborating evidence such as spraints or otter prints is usually needed to be sure it is an otter slide.

Spraint: faecal scent marks left by otters, often at strategic features.

Spraint(ing) site: this could be any site where otters deposit spraint. Many are used habituality, and thus regular spraint sites may have many spraints present of different ages and states of weathering. They may be more visible because of changes in vegetation, such as greener tufts or a halo of yellow vegetation.

Swapping out: in the context of camera-trapping, this is the process of swapping the whole camera-trap for a different one (with fresh batteries and SD card) at an emplacement, as opposed to changing the batteries and SD card of a camera and redeploying it. This can save time during maintenance visits and reduce time spent at sensitive sites.

Time-lapse: a setting that some camera-trap models have that allows images to be taken at programmed time intervals rather than relying on a triggering mechanism.

Trigger delay: this is a programmable time delay between a camera-trap taking an image or finishing recording a video clip and it being rearmed and ready to retrigger.

Trigger time: often also called trigger speed, this is the time delay between the PIR detector of a camera being triggered by an animal and the camera-trap taking an image or beginning to record a video clip.

Trigger probability: in the context of camera-trapping, the probability that a camera-trap's PIR sensor is triggered by an animal, regardless of whether this produces usable video clips or images. It is useful to separate this component of detection probability because it is influenced by different factors to registration probability.

Walk-over survey: a basic survey whereby the aim is to identify presence of otter on site and/or to look for field signs that indicate otter activity patterns and identify resting places or potential resting places.

Weaning: the process of a female mammal gradually introducing her young to what will be their adult diet, while withdrawing the supply of milk. In otters, this is thought to begin at eight to ten weeks old.

Zone of influence/impact: the area over which a development will have ecological impacts. Its size depends on the site's features and the nature of the proposed development.

References

Apps, P.J., McNutt, J.W., 2018. How camera traps work and how to work them. *African Journal of Ecology* 56, 702–709. https://doi.org/10.1111/aje.12563

Chanin, P., 2003a. Monitoring the Otter *Lutra lutra*. Conserving Natura 2000 Rivers Monitoring Series No. 10. English Nature, Peterborough.

Chanin, P., 2003b. Ecology of the European otter: *Lutra Lutra*. Life in UK Rivers, Peterborough.

Chanin, P., 1985. *The Natural History of Otters*. Christopher Helm mammal series. Helm, London.

Chanin, P.R.F., 1993. *Otters*, British natural history series. Whittet, London.

Clark, M., 1992. *Badgers*. Whittet Books, London.

Coxon, J., 2017. Investigating the variables affecting camera-trap detectability of two semi-aquatic mammals: the Eurasian beaver (*Castor fiber*) and the Asian small-clawed otter (*Aonyx cinerea*) (Unpublished MSc Thesis). Edinburgh Napier University.

Danes, E., 2024. Developing a novel methodology to estimate the age of otter (*Lutra lutra*) cubs from camera trap footage using proportional body measurements (Unpublished MSc Thesis). Edinburgh Napier University.

Dean, M., 2021. *Water Vole Field Signs and Habitat Assessment: A Practical Guide to Water Vole Surveys*. Conservation Handbooks. Pelagic Publishing, Exeter.

Dugdale, H.L., Macdonald, D.W., Pope, L.C., Johnson, P.J., Burke, T., 2008. Reproductive skew and relatedness in social groups of European badgers, *Meles meles*. *Molecular Ecology* 17, 1815–1827. https://doi.org/10.1111/j.1365-294X.2008.03708.x

Duggan, L., 2016. An investigation of possible sources of bias in the detection probability of Asian short-clawed otters (*Aonyx cinereus*) (Unpublished MSc Thesis). Edinburgh Napier University.

Durbin, L.S., 1996. Some changes in the habitat use of a free-ranging female otter *Lutra lutra* during breeding. *Journal of Zoology* 240, 761–764. https://doi.org/10.1111/j.1469-7998.1996.tb05319.x

European Commission, 2021. Guidance document on the strict protection of animal species of community interest under the Habitats Directive: a summary. European Commission, Brussels.

European Commission, 2007. Guidance document on the strict protection of animal species of Community interest under the Habitats Directive 92/43/EEC. European Commission, Brussels.

Findlay, M., Alexander, L., Macleod, C., 2015. Site condition monitoring for otters (*Lutra lutra*) in 2011-12. Scottish Natural Heritage Commissioned Report No. 521. Scottish Natural Heritage, Inverness.

Findlay, M.A., Briers, R.A., Diamond, N., White, P.J.C., 2017. Developing an empirical approach to optimal camera-trap deployment at mammal resting sites: evidence from a longitudinal study of an otter *Lutra lutra* holt. *European Journal of Wildlife Research* 63, 96. https://doi.org/10.1007/s10344-017-1143-0

Findlay, M.A., Briers, R.A., Ingledew, R.P., White, P.J.C., 2023. An evidence-based approach to identifying resting sites of Eurasian otter *Lutra lutra* from camera-trap and field-sign data. *Wildlife Biology*, e01036. https://doi.org/10.1002/wlb3.01036

Findlay, M.A., Briers, R.A., White, P.J.C., 2020. Component processes of detection probability in camera-trap studies: understanding the occurrence of false-negatives. *Mammal Research* 65, 167–180. https://doi.org/10.1007/s13364-020-00478-y

Garcia De Leaniz, C., Forman, D.W., Davies, S., Thomson, A., 2006. Non-intrusive monitoring of otters (*Lutra lutra*) using infrared technology. *Journal of Zoology* 270, 577–584. https://doi.org/10.1111/j.1469-7998.2006.00124.x

Green, J., Green, R., Jeffries, D.J., 1984. A radio-tracking survey of otters *Lutra lutra* on a Perthshire river system. *Lutra* 27, 85–145.

Hauer, S., Ansorge, H., Zinke, O., 2002. Reproductive performance of otters *Lutra lutra* (Linnaeus 1758) in Eastern Germany: low reproduction in a long-term strategy. *Biological Journal of the Linnean Society* 77, 329–340. https://doi.org/10.1046/j.1095-8312.2002.00097.x

Heap, C.J., Lafontaine, L., Field, D., 2010. Summary of husbandry guidelines for the Eurasian otter in captivity. IUCN/SSC Otter Specialist Group, Otters in Captivity Task Force.

Heggberget, T.M., Christensen, H., 1994. Reproductive timing in Eurasian otters on the coast of Norway. *Ecography* 17, 339–348. https://doi.org/10.1111/j.1600-0587.1994.tb00111.x

Hong, S., Di Febbraro, M., Kim, H.G., Loy, A., 2022. Is scat marking a reliable tool for otter census and surveys at the landscape scale? *Journal of Environmental Management* 315, 115098. https://doi.org/10.1016/j.jenvman.2022.115098

Kean, E.F., Chadwick, E.A., Müller, C.T., 2015. Scent signals individual identity and country of origin in otters. *Mammalian Biology* 80, 99–105. https://doi.org/10.1016/j.mambio.2014.12.004

Kirstner, F., Slaney, L., Jewell, Z., Ben-David, A., Alibhai, S., 2022. It's otterly confusing – distinguishing between footprints of three of the four sympatric Asian otter species using morphometrics and machine learning. *OTTER, Journal of the International Otter Survival Fund* 108–125.

Kruuk, H., Carss, D.N., Conroy, J.W.H., Gaywood, M.J., 1998. Habitat use and conservation of otters (*Lutra lutra*) in Britain: a review, in: *Behaviour and Ecology of Riparian Mammals*. Cambridge University Press, Cambridge, pp. 119–134.

Kruuk, H., Conroy, J.W.H., Glimmerveen, U., Ouwerkerk, E.J., 1986. The use of spraints to survey populations of otters *Lutra lutra*. *Biological Conservation* 35, 187–194. https://doi.org/10.1016/0006-3207(86)90050-9

Kruuk, H., Moorhouse, A., 1990. Seasonal and spatial differences in food selection by otters *(Lutra lutra)* in Shetland. *Journal of Zoology* 221, 621–637. https://doi.org/10.1111/j.1469-7998.1990.tb04021.x

Lerone, L., Carpaneto, G.M., Loy, A., 2015. Why camera traps fail to detect a semi-aquatic mammal: Activation devices as possible cause. *Wildlife Society Bulletin* 39, 193–196. https://doi.org/10.1002/wsb.508

Liles, G., 2003. Otter Breeding Sites Conservation and Management. Conserving Natura 2000 Rivers Conservation Techniques Series No. 5. English Nature, Peterborough.

Martin-Walker, A., 2017. What influences camera trap detectability of Eurasian Beavers *Castor fiber* and Asian short clawed otters *Aonyx cinereus*? (Unpublished MSc Thesis). Edinburgh Napier University.

Meek, P.D., Ballard, G.-A., Fleming, P.J.S., Schaefer, M., Williams, W., Falzon, G., 2014. Camera Traps Can Be Heard and Seen by Animals. *PLoS ONE* 9, e110832. https://doi.org/10.1371/journal.pone.0110832

Mercier, L., Fried, G., 2004. Preliminary Study of the Tracks of Captive Otters (*Lutra lutra*) as a Tool for Field Research. *IUCN Otter Specialist Group Bulletin* 21, 94–99.

Moorhouse, A., 1988. Distribution of holts and their utilisation by the European otter (*Lutra lutra* L.) in a marine environment (Unpublished MSc Thesis). University of Aberdeen.

NatureScot, 2024. Standing advice for planning consultations – Otters. NatureScot, Inverness.

Oliver, K., 2023. Investigating Seasonality of Otter Breeding in Britain Using Newspaper articles, Camera Trap Footage and Photograph Data (Unpublished MSc Thesis). Edinburgh Napier University.

Randler, C., Kalb, N., 2018. Distance and size matters: A comparison of six wildlife camera traps and their usefulness for wild birds. *Ecology and Evolution* 8, 7151–7163. https://doi.org/10.1002/ece3.4240

Reid, N., Lundy, M.G., Hayden, B., Lynn, D., Marnell, F., McDonald, R.A., Montgomery, W.I., 2013. Detecting detectability: identifying and correcting bias in binary wildlife surveys demonstrates their potential impact on conservation assessments. *European Journal of Wildlife Research* 59, 869–879. https://doi.org/10.1007/s10344-013-0741-8

Robitaille, J.-F., Proulx, G., Do Linh San, E., 2020. On the Use of Den and Resting Site Terminology for Species in the *Martes* Complex. *Canadian Wildlife Biology & Management* 9, 137–151.

Roche, E., 2016. Detection probability of asiatic small-clawed otter (*Aonyx cinereus*) using camera trapping technique at Edinburgh Zoo (Scotland) (Unpublished MSc Thesis). Edinburgh Napier University.

Romanowski, J., Brzeziñski, M., 1997. How standard is the standard technique of the otter survey? *IUCN Otter Specialist Group Bulletin* 14, 57–60.

Rosoux, R., Libois, R.M., 1996. Use of day resting places by the European otter (*Lutra lutra*) in teh Marais Poitevin (France). A radiotracking study, in: *Proceedings of the 1st European Congress of Mammology*. Lisboa, pp. 119–212.

Taylor, P.S., Kruuk, H., 1990. A record of an otter (*Lutra lutra*) natal den. Journal of Zoology 222, 689–692. https://doi.org/10.1111/j.1469-7998.1990.tb06024.x

Van Berkel, T., 2014. *Camera trapping for wildlife conservation: expedition field techniques*. Geography Outdoors, London.

Wayre, P., 1979. *The Private Life of the Otter*. Batsford, London.

Wege, D., 2024. *Mammal tracks of Europe*. David Wege Nature, Cambridge.

Weinberger, I.C., 2016. The Eurasian Otter (*Lutra lutra*) in the Alpine Arc: Resource selection and habitat suitability models (Unpublished PhD thesis). Universität Zürich, Zurich.

Weinberger, I.C., Muff, S., Kranz, A., Bontadina, F., 2019. Riparian vegetation provides crucial shelter for resting otters in a human-dominated landscape. *Mammalian Biology* 98, 179–187. https://doi.org/10.1016/j.mambio.2019.09.001

Welbourne, D.J., Claridge, A.W., Paull, D.J., Lambert, A., 2016. How do passive infrared triggered camera traps operate and why does it matter? Breaking down common misconceptions. *Remote Sensing in Ecology & Conservation* 2, 77–83. https://doi.org/10.1002/rse2.20

Wellington, K., Bottom, C., Merrill, C., Litvaitis, J.A., 2014. Identifying performance differences among trail cameras used to monitor forest mammals. *Wildlife Society Bulletin* 38, 634–638. https://doi.org/10.1002/wsb.425

White, P.J.C., Findlay, M.A., 2024. Do UK survey licenses for otter support best practice? CIEEM. *In Practice* 123, 48–52.

Yoxon, P., Yoxon, K., 2014. Estimating Otter Numbers Using Spraints: Is It Possible? *Journal of Marine Biology* 2014, 1–6. https://doi.org/10.1155/2014/430683

Index

- activity patterns 113–15
- age of emerged cubs 98
- American mink 24, 115, 116, 117
- angle, camera-trap 89–92
- artificial holts 100
- artificial intelligence (AI), use of 112–13
- Austrian radio tracking study 62

- bedding and nests 44–8, 69, 124
- behaviours 124–5
  - bedding collection 124, 126, 137
  - events list, adding to 125
  - latrine 124
  - other 125
  - stationary 124–5
- bird droppings 12, 13
- black IR 84, 87, 88
- breeding, definition of 3
- breeding sites/places 39–41
  - maternal resting places 49
  - natal resting places 48–9

- camera-trapping of couches 94–5
- camera-trapping licence 82
- camera-trapping research 42–3
- camera-traps
  - checklist, pre-deployment 104–5
    - construction, monitoring otter 'activity' during 129
  - deployment of 82–109
  - detection probability 76–80
  - field of view of most camera-trap models 73–4
  - illumination 74–5
  - ledges and culverts 130
  - monitoring artificial resting places 129–30
  - multiple-trap array 89
  - non-resting places 129
  - passive infrared detection 72–3
  - terrestrial routes 128–9
- capture quality 79–80
- Chanin, Paul 2
- checklist, camera-trap 104–5
- coastal sites 33–4
- code of conduct 83
- construction, monitoring otters during 99–100
- core range 97
- couches 2–4, 19, 36, 37, 41, 52, 94–5

- data management 104
- dense vegetation 92–4
- detection probability 76–80
- detection zone 73
- determining age 121–2
- determining sex 117–21
- distance, camera-trap 89–92
- disturbance 6–7
- disturbance, resting place 51–2
- duration, camera-trap 96–7

- emplacement, camera-trap 92
- entrance, resting place 49–51

- false trigger 76, 85, 97, 100, 103, 107, 112, 114, 124, 132
- faults, camera-trap surveys 107–8
- field of view 66, 73–80, 85, 92, 95, 99–104, 113, 115, 131, 133
- field signs
  - coastal 33–4
  - field evidence 30–1
  - food remains 26–9
  - footprints 22–5
  - freshwater 31–2
  - hairs 30
  - latrine 15–18
  - paths and slides 25–6
  - rainfall 30
  - sign heaps 18–19

in site surveys 34–5
spraint sites 14–15
spraints 9–14
washing pools and rolling pitches 19–22
field surveys, expectations of 41–3
food remains 26–9
footage, preparing 112–13
footprints 22–5
American mink *vs.* otter 24
dog *vs.* otter 23
freshwater habitats 19–22, 26, 31–2
freshwater sites 31–2

gallery, resting place 52–61
groups of otters 114–15

Habitats Directive 4, 38
Habitats Regulations 4, 36
hairs, as field sign 30
haul-out sites 3
health and safety 7
height, camera-trap 89–92
hidden latrines 44
holts 2–4, 31, 32, 34, 36–42, 44, 48–52, 54, 56, 57, 59, 64, 66, 68–70, 76, 77, 80, 86, 87, 89–106, 114, 115, 117, 120, 121, 124, 126–35
hovers 3

individual otters, identification of 122–3
infrared (IR) radiation 72, 74, 75, 77, 87, 91
infrared LEDs 73–5
Internal Drainage Board 31
interpreting camera-trap data
activity patterns 113–15
behaviours 124–5
camera-traps, scenarios 128–30
footage, preparing 112–13
natal and maternal resting places, characteristics of 125–8
otter information 115–24
quality control 130–4
reporting, summarising/interpreting your data 134–7
resting places, types of activity 125–8
survey effort 111–12
invasive surveys 64–7
potential disturbance 66
how to undertake 66
what to look for in 66–7
inverse square law 77

latrines 15–18, 45, 52, 54, 57–60, 66–68, 69
drier pellet-type droppings 15, 17
enriched by badger dung 16, 18
maternal resting 49
resting places 44
vs. spraint pile 18, 19
legislation 4, 38
licensing 4, 6–7
litter 50, 94, 97, 121, 122, 126, 127
'low-glow' LEDs 74–5, 101

maintenance visits 103–4
maternal holt 2, 3, 32, 40, 54, 56, 57, 59, 127
maternal resting places 49
characteristics of 125–8
mill lade 53
mink scats 12, 13

natal holt 39, 48–9, 97
sharing of 127
natal resting places 48–9
characteristics of 125–8
Nature Conservancy Council 2
'no-glow' LEDs 75, 101

paired event, otters 113–14
passive infrared (PIR) detection 72–3, 91
sensitivity of 85
paths and slides 25–6
paths, camera-trap 98–9
potential resting places 61–4
prepuce 118–19
prey remains
see food remains
problematic sites
complex habitat 106–7
entrances close to water 104–6
reedbeds 107
underwater entrances 106

radio-tracking research 42–3, 51
red IR 87, 88, 101
reedbeds, otters and 64, 107

registration probability 79
resting places 2–3, 38
  bedding and nests 44–8
  camera-trap, best pattern and timing to 95–7
  case study 68–70
  decision tree 67–8
  disturbance 51–2
  entrances 49–51
  gallery 52–61
  latrines 44
  multiple camera-traps 89
  sleeping areas 51
  surrounding habitat 52
  types of activity 125–8
rolling pitch 19–22, 26, 30, 95, 99

set up, camera-trap 101–2
sign heaps 18–20, 26
site surveys 4–6
sleeping area 51
sniffing etiquette, spraint 14
spraints
  coastal 9, 11
  definition of 9
  dried black 9, 10
  'mash' form 9, 10
  sites 14–15, 98
  size 9
  sniffing etiquette 14
  variety of, on single site 12
  vs. bird droppings 12–14
  vs. mink scats 12–14
  weathered 11, 12
still images *vs.* video footage 85–6
survey methods 2, 5
  see also specific methods

temporary resting sites 3–4
terminology 2–4, 36
theft/damage, protecting camera-traps 94
time-lapse camera-trap 95, 107
trigger delay 76, 86
trigger probability 77–9, 90–1, 99
trigger time 76, 85–6, 101

urination 113, 118, 121

video clips 86

walk-over surveys
  bedding and nests 44–8
  latrines 44
  maternal resting places 49
  natal resting places 48–9
washing pools 19–22
water vole 2, 32
weaning period 39–40
weathered spraint site 11, 12
Wildlife and Countryside Act (1981) 4, 38

zone of influence/impact 31, 33